PERGAMON INTERNATIONAL LIBRARY
of Science, Technology, Engineering and Social Studies
The 1000-volume original paperback library in aid of education,
industrial training and the enjoyment of leisure
Publisher: Robert Maxwell, M.C.

INTERNATIONAL SERIES ON
MATERIALS SCIENCE AND TECHNOLOGY
VOLUME 17—EDITOR: D. W. HOPKINS, M.Sc.

The Fundamentals of Corrosion

Other Titles in the International Series on
MATERIALS SCIENCE AND TECHNOLOGY

The Fundamentals
of Corrosion

J. C. SCULLY, M.A., PH.D., A.I.M.
Reader in Corrosion Science at the University of Leeds

SECOND EDITION

PERGAMON PRESS
OXFORD · NEW YORK · TORONTO
SYDNEY · PARIS · FRANKFURT

U.K.	Pergamon Press Ltd., Headington Hill Hall, Oxford OX3 0BW, England
U.S.A.	Pergamon Press Inc., Maxwell House, Fairview Park, Elmsford, New York 10523, U.S.A
CANADA	Pergamon of Canada Ltd., 75 The East Mall, Toronto, Ontario, Canada
AUSTRALIA	Pergamon Press (Aust.) Pty. Ltd., 19a Boundary Street, Rushcutters Bay, N.S.W. 2011, Australia
FRANCE	Pergamon Press SARL, 24 rue des Ecoles, 75240 Paris, Cedex 05, France
FEDERAL REPUBLIC OF GERMANY	Pergamon Press GmbH, 6242 Kronberg/Taunus, Pferdstrasse 1, Federal Republic of Germany

First edition 1966
Second edition 1975
Reprinted (with minor corrections) 1978

Library of Congress Catalog Card No. 74-28581

.

Printed in Great Britain by William Clowes & Sons Limited, London, Beccles and Colchester

ISBN 0 08 018081 7 (hardcover)
ISBN 0 08 018080 9 (flexicover)

To my wife Celia

Contents

4. Corrosion Failures and Attack

Introduction

CORROSION can be defined as the reaction of a metallic material with its environment. The products of this reaction may be solid, liquid or gaseous. Both the physical and chemical natures of the products are important since they frequently influence the subsequent rate of reaction.

The study of corrosion needs neither justification nor explanation. Every metal producer and user is forced to undertake it. The amount of money spent in an industrial country in combating corrosion by preventative measures, e.g. painting or plating, by replacement of corroded parts, by the use of expensive alloys, etc., is extremely high. Estimates of this sum, just for Great Britain alone, come to £1365 million per annum. The corrosion of metals, therefore, represents a terrible waste of both natural resources and money.

Corrosion studies are undertaken by all types of physical scientist and any investigator needs to understand the language of each if he is to make full use of previous work and to integrate this with his own. Although the semiconductivity of oxides, electrochemical kinetics and surface dislocation arrangements are usually studied by physicists, electrochemists and metallurgists respectively, all three are of considerable importance in studying corrosion reactions. The outlook of the corrosion scientist, quite independent of his training, must need be truly interdisciplinary and this breadth helps make corrosion a particularly fascinating subject.

It is hoped that this book will be readily suitable for students at all levels. It originates from the lecture course on corrosion which is given by me in this university, but it is written so that it is suitable for students and other readers who have only an elementary knowledge of physics and chemistry. It is intended for all who desire to understand the fundamentals of this pernicious phenomenon. In this book the basic theories on corrosion and their application to its prevention are presented in three chapters, while a fourth describes the types of failure and attack that occur and the book concludes with a few miscellaneous topics. Reference is made throughout to the work of many investigators and a complete list of these is given at the end of the book. It is my earnest hope that I have omitted no warranted reference. A bibliography has also been provided which is divided into four sections. These contain lists of other introductory books, general and advanced books, specialized books and pure science books which cover topics that are only lightly touched upon in this book, e.g. chemical thermodynamics.

The first edition of this book appeared early in 1966. In the last 8 years there has developed an increasing awareness that the earth does not represent a boundless cornucopia. Instead it has resources that must be husbanded since once lost they may not be recoverable and their replacement will involve continuously increasing expense. The study of corrosion has a bearing on all such broad considerations.

In preparing the second edition of this book I have been able to expand some sections, bring others up-to-date, delete others, tidy parts up and correct the occasional infelicities that crept into the first edition despite my vigilance. Inevitably the book has grown in size (about 20% larger). Apart from these changes, however, the same pattern has been retained as in the first edition.

Many of the diagrams in this book have been redrawn from published books and scientific papers. I am indebted to the authors and publishers concerned who have given their permission for this and for the reproduction of tables. I wish to acknowledge the following sources: American Society of Metals, Table II and Fig. 87; E. Arnold, Figs. 21, 45 and Tables V and IX; Butterworths, Figs. 21, 25, 26, 27 and Tables I and III; Cambridge University Press, Fig. 1; Chapman & Hall, Table X; Electrochemical Society, Figs. 20, 47, 57, 58, 65, 67, 75, and Table VII; Faraday Society, Fig. 6; Garnat Mather, Fig. 2; Gauthier-Villars, Figs. 31, 32, 33; *Industrial and Engineering Chemistry*, Fig. 59; Institute of Metals, Figs. 7, 13; Iron and Steel Institute, Figs. 15, 18, 19, 22; *Metal Industry*, Fig. 23; Metallurgical Society of AIME, Fig. 5; McGraw-Hill, Fig. 17; National Association of Corrosion Engineers, Fig. 86; Newnes, Fig. 66; Pergamon Press, Figs. 74, 80, 88, 89; Prentice-Hall, Table VI; Royal Society, Figs. 24, 63; Society for Chemical Industry, Figs. 68, 69; Sociétés Chimiques Belges, Fig. 14; *Trans. Chim. Pays-Bas*, Fig. 46; Van Nostrand, Table VIII; Verlag Chemie, Fig. 3.

The preparation of a book requires much work. I am grateful to Mrs. J. C. Nathan for patiently and intelligently typing the manuscript and to Messrs. R. Trippitt and G. Dean for skilfully drawing the diagrams. I have also much appreciated the generous and helpful comments of the General Editor, Mr. D. W. Hopkins of the University College of Swansea.

Finally I continue to remain indebted to those who have taught me, particularly Dr. T. P. Hoar, and to those whom I worked with some years ago, notably Dr. C. Edeleanu and Dr. M. J. Pryor.

J. C. SCULLY

Department of Metallurgy
Houldsworth School of Applied Science,
The University,
Leeds LS2 9JT, Yorkshire

CHAPTER 1

Oxidation

1.1. INTRODUCTION

When any metal combines with an atom or with a molecular group and loses electrons, then an oxidation reaction has taken place. A metal is also oxidized and loses electrons when it goes from one valency to a higher one. The term oxidation, therefore, describes the transfer of electrons, and reactions involving oxygen combining with metals form only a small section under the general heading. In this chapter, however, all the discussion is concerned specifically with the formation of metal oxides.

When a metal is oxidized, the other species taking part in the reaction is reduced, i.e. it gains electrons. Some common examples can be cited:

Oxidized species		Reduced species
4 Cu	$+O_2 =$	2 Cu_2O
2 Cu_2O	$+O_2 =$	4 CuO
2 Fe	$+O_2 =$	2 FeO

The overall reaction for an oxidation of this type can be represented as two separate reactions occurring simultaneously:

Oxidation: Me → $Me^{z+} + z$ electrons, where z is the valency of the metal.

1

Reduction: $O_2 + z$ electrons $\rightarrow \dfrac{z}{2} O^{2-}$.

Combined: $Me + O_2 \rightarrow MeO_{z/2}$.

It is reasonable to assume that a univalent metal would form only one oxide, whereas a multivalent metal would form as many different oxides as there are valencies. During oxidation processes, however, it is not unknown for metals to exhibit, either temporarily or permanently, unusual valencies, so that the number of valencies possessed by a metal will not always correspond to the normal number of different oxidized states. In addition, it is frequently found that any oxide may have more than one structure, a property referred to as polymorphism. An example of a polymorphic oxide would be TiO_2, titania, which can exist as rutile, anatase and brookite. These have crystallographic structures that are tetragonal (the first two with different lattice parameters) and orthorhombic, respectively. Rutile is the normal form of titania.

Electron donation and acceptance occur in several types of chemical bonding. Metal oxides, sulphides, etc., exhibit predominantly ionic bonding. A metal oxide consists of positively charged metal ions, Me^{z+}, and larger negatively charged oxygen ions, O^{2-}. The sum total of all the positive charges is equal to the sum total of all the negative charges and an oxide is thereby electrically neutral. Every oxide has a definite crystallographic structure and the two types of ions are distrib-

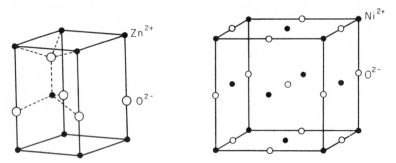

Fig. 1. Atomic models[1] of two different oxides: ZnO and NiO.

uted within this at different specific sites. Two oxide structures are illustrated in Fig. 1. It is often convenient to consider the metal and oxygen ions as existing in two separate but interpenetrating lattices and for each of these the term sublattice is used. Where two different oxides have the same crystallographic structure, the size of the smallest repeat unit of the lattice, which is called the lattice parameter, will be different for each. They can therefore be distinguished from each other by several methods, e.g. X-ray diffraction.

Oxides are composed of grains that exhibit behaviour similar to that of a metal. An oxide can recrystallize, exhibit grain growth and it may deform plastically, particularly at high temperatures. Diffusion rates will be higher along the intercrystalline paths than within the oxide crystals. Similarly, diffusion rates on the surfaces of crystals will be greater than within the bulk. While it is realistic to discuss growth by consideration of the kinetics of a thickening parallel-sided layer, as is done in section 1.3, on a localized scale growth may be somewhat irregular and give rise to a wide variety of outer oxide surface morphology.

1.2 THERMODYNAMICS OF OXIDATION

When a metal oxidizes, there is a change in the free energy, G, of the system which is equal to the work done or absorbed during the process. It is a maximum when the process takes place reversibly. It is this change in the free energy of the system that is the driving force of the reaction and it represents the maximum fraction of the energy that can be converted into work. The performance of this work must be accompanied by a decrease in the free energy of the system, $(-\Delta G)$, otherwise the reaction cannot take place.

The change in free energy, ΔG, can be represented as:

$$\Delta G = G \text{ (products)} - G \text{ (reactants)}. \tag{1}$$

Sometimes the term free energy is also used to describe the Helmholtz function or work function which is denoted by the symbol F. The two are related by $G = F + pV$. Some authors describe the Gibbs free energy by the symbol F. All discussion in this book on free energy is confined to the Gibbs function, G.

The standard free energy change for the formation of nearly all metal oxides is negative, i.e. oxides are thermodynamically stable in oxygen atmospheres whereas metals are not. Oxidation will therefore always tend to occur.

For the reaction $Me + O_2 \rightleftharpoons MeO_2$, the equilibrium constant, K, derived from the law of mass action, is equal to $[MeO_2]/[Me][O_2]$ where the species within the rectangular brackets represent the active masses of the reacting substances. The active masses of the solid metal and oxide are taken by convention as equal to unity and the active mass of the oxygen can be represented by its partial pressure under equilibrium conditions. If the oxygen pressure is in atmospheres, then the new equilibrium constant, K_p, is equal to $1/p_{O_2}$.

The equilibrium constant of a reaction is related to the free energy change. This can be shown very simply by considering a mixture of four ideal gases A, B, C and D, where an ideal gas is defined as obeying the relationship $PV = RT$. These are put into a chamber and then react according to:

$$A + B \rightleftharpoons C + D.$$

In general, the work done in changing a volume of a gas from V_1 to V_2 is equal to

$$\int_{V_1}^{V_2} p.dV = RT \int_{V_1}^{V_2} \frac{1}{V} dV = RT \ln \frac{V_2}{V_1} = RT \ln \frac{P_1}{P_2}.$$

If the four gases are at initial pressures of P_A^1, P_B^1, P_C^1 and P_D^1 and at final equilibrium pressures of P_A^2, P_B^2, P_C^2 and P_D^2, then the work done in reaching equilibrium is

$$RT \ln \frac{P_A^1 P_B^1}{P_A^2 P_B^2} + RT \ln \frac{P_C^2 P_D^2}{P_C^1 P_D^1}$$

and this is equal to $-\Delta G$.

This can be rearranged:

$$-\Delta G = RT \ln \frac{P_C^2 P_D^2}{P_A^2 P_B^2} + RT \ln \frac{P_A^1 P_B^1}{P_C^1 P_D^1}.$$

But

$$\frac{P_C^2 P_D^2}{P_A^2 P_B^2} = K_p,$$

the equilibrium constant.

Therefore, the free energy change of the reaction, ΔG, is related to the equilibrium constant K_p by the equation:

$$\Delta G = -RT \ln K_p + RT \Sigma n \ln p \qquad (2)$$

where the term $RT \Sigma n \ln p$ defines the initial and final states of the system, n and p representing respectively the number of moles and the pressure of the participants in the reaction, and the terms $n \ln p$ being added in the same sense as in equation (1). Unlike the equilibrium constant these terms are variable. If the pressure of the oxygen is atmospheric, $RT \Sigma n \ln p$ for a simple oxidation reaction will be equal to zero. Then:

$$\Delta G^\circ = -RT \ln K_p \qquad (3)$$

where ΔG° is defined as the standard free energy change of the reaction.

When any chemical reaction occurs, the masses of the reactants and products are decreasing and increasing respectively. Since the internal energy of a substance is directly related to the number of molecules of that substance, that of the reactants will decrease and that of the products increase. The term, chemical potential, represented by μ, is used to denote the change of free energy of a substance with change in the number of moles, n, of a substance in a reaction in which temperature, pressure and the numbers of moles of all other substances are constant. Thus:

$$\mu = \mu^\circ + RT \ln a \qquad (4)$$

where $a =$ the activity of the material (defined on p. 68) and μ° the chemical potential of one mole at unit activity.

The free energy change of the oxidation reaction $Me + O_2 \rightarrow MeO_2$ is equal to the arithmetical difference of the chemical potentials of all the phases present.

For the metal: $\mu_{Me} = \mu^{\circ}_{Me} + RT \ln a_{Me}.$

For the oxygen: $\mu_{O_2} = \mu^{\circ}_{O_2} + RT \ln a_{O_2}.$

For the oxide: $\mu_{MeO_2} = \mu^{\circ}_{MeO_2} + RT \ln a_{MeO_2}.$

The free energy change, ΔG, for the oxidation reaction is equal to $\mu_{MeO_2} - \mu_{Me} - \mu_{O_2}$ and $\mu^{\circ}_{MeO_2} - \mu^{\circ}_{Me} - \mu^{\circ}_{O_2} = \Delta G^{\circ}$ for the reaction. Since the activities of the metal and metal oxide are equal to unity,

$$\Delta G = \mu^{\circ}_{MeO_2} - \mu^{\circ}_{Me} - \mu^{\circ}_{O_2} - RT \ln a_{O_2} = \Delta G^{\circ} - RT \ln p_{O_2}.$$

But $\Delta G^{\circ} = -RT \ln K_p.$

Therefore $\Delta G = -RT \ln K_p - RT \ln p_{O_2}$

$$= -RT \ln \frac{1}{p'_{O_2}} - RT \ln p''_{O_2} \qquad (5)$$

where p'_{O_2} = oxygen pressure at equilibrium and p''_{O_2} = the initial oxygen pressure at the very beginning of the reaction.

$\Delta G = 0$ when the initial pressure of oxygen corresponds to the partial pressure of oxygen as represented in the equilibrium constant. Under these conditions there is no driving force for the reaction, and the oxide and metal are then equally stable. If the experimental pressure is lowered below this value the oxide will dissociate. This critical value of pressure, which varies with temperature, is called the *dissociation pressure* of the oxide. If several oxides are formed on a metal, e.g. Fe_2O_3, Fe_3O_4 and FeO, they will have different dissociation pressures and the oxide that is richest in oxygen will usually dissociate to an oxide containing less oxygen and not to bare metal directly.

For most metal oxides the values of oxygen partial pressure that are required for dissociation are too low to be obtained experimentally. In the case of gold, however, ordinary atmospheric oxygen (0·2 atm) is too low for a stable oxide at room temperature. From equation (5), however, raising the initial pressure of oxygen well above atmospheric would make the free energy change less positive and eventually negative, rendering the oxide stable. Under ordinary dry atmospheric conditions a gold surface is not actually bare since it is covered with a layer of chemisorbed oxygen atoms. For silver the free energy change of formation of the oxide is negative at room temperature but becomes zero at 200 °C. If silver is heated gently, the oxide will dissociate at that temperature. For some other metals, e.g. copper and nickel, the partial pressures required for oxide dissociation are obtainable experimentally.

The values of free energy change for some metal oxides are shown in Table I[2].

TABLE I. STANDARD FREE ENERGY $(-\Delta G^\circ)$ OF FORMATION OF SOME OXIDES PER MOLE OF OXYGEN AT 300K[2]

	kcal	kJ
Ag_2O	5·1	21·3
Cu_2O	69·2	289·2
PbO	90	376·2
NiO	102·8	430
FeO	109·2 (at 500K)	455·6
ZnO	152·4	631
MgO	273	1140
SiO_2 (quartz)		824
Cr_2O_3	157·8 (at 500K)	660
Al_2O_3	252	1060

The standard change in free energy, ΔG°, in kJ/mole, is related to ΔH°, the standard heat of the reaction, in the same units, ΔS°, the standard change in entropy, and T, the absolute temperature, by the equation:

$$\Delta G^\circ = \Delta H^\circ - T\Delta S^\circ. \qquad (6)$$

The numerical value of the free energy change for oxidation reactions decreases with increase in temperature, as shown in Fig. 2[3].

The Fundamentals of Corrosion

Physically, this is readily interpretable. If a *solid* metal, Me, reacts with a *gas*, oxygen, and forms a stable *solid* oxide, MeO, then the disorder of that system, measured by the change in entropy, is decreased. The reaction solid + gas → solid produces a more regular

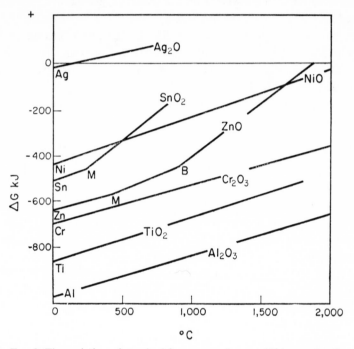

FIG. 2. The variation of standard free energy change, $\Delta G°$/per mole of oxygen, with temperature for a number of metal oxides[3].

atomic arrangement of the oxygen atoms in the oxide than that previously existing in the gaseous state. For this reaction, therefore, ΔS is negative and the slope of ΔG vs. T positive. This is true for all metal oxides. If the melting point of the metal is exceeded the slope $\Delta G/T$ is increased slightly since the reaction liquid metal + gaseous oxygen → solid oxide produces a slightly greater decrease in entropy. The reverse is true when the melting point of the oxide is exceeded.

Although considerations of free energy changes indicate the more stable reaction product, they cannot predict the final outcome of a reaction. Thermodynamic factors do not include kinetic parameters and these are of paramount importance. Since the free energy changes for the formation of most metal oxides are negative, all metal should revert to a combined state when exposed to the atmosphere. It is largely due to the kinetics of oxidation reactions that this does not occur. The formation of a metal oxide on a bare metal surface usually restricts further access of one reactant to the other. When a metal is abraded at room temperature, the existing oxide is broken, but it very rapidly reforms on any bare surface although growth quickly stops after a certain thickness, 1–4 nm (10–40 Å), is attained. Unless this occurs, the metal will be entirely consumed to form oxide. Thus it must be emphasized that the very existence of a metal is dependent upon the naturally formed protective oxide skin. Over long periods of time, millions of years, most metals will be converted to a combined state with oxygen or other reactive species provided that these are present in sufficient quantities. Only those metals with low free energies of oxide formation are likely to occur even in the free state. Copper and gold are examples of such metals.

Some interesting experiments on the oxidation of copper[4] indicate the important distinction between thermodynamic and kinetic factors. Copper was oxidized at different temperatures and partial pressures of oxygen. The results are shown in Fig. 3. If the pressure during an experiment was increased from below the dissociation pressure, the metal surface was covered with an extremely thin layer of adsorbed oxygen (possibly one layer thick). Subsequently oxide nuclei formed. These nuclei did not cover the surface entirely at first, since time was required for their lateral growth. During this period of time that part of the metal surface which was covered only with the very thin layer was thermodynamically unstable with respect to oxygen, but it was the kinetic factors that were rate-controlling. At higher pressures the increased concentration of oxygen produced a very high nucleation rate and the metal was entirely covered with a thicker oxide almost instantaneously.

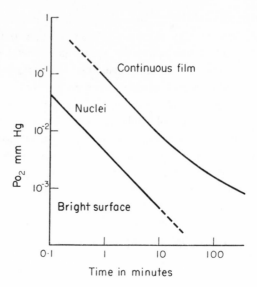

FIG. 3. The effect of variation in oxygen partial pressure upon the nucleation of oxide on a copper surface[4].

Nuclei often exhibit epitaxial relationships with the metal surface on which they form. This can be explained energetically by the most favourable movements of the metal atoms in converting from the close packing of a metal lattice to the more open packing of the metal ions in an oxide structure in which the large space is occupied by the oxygen ions. Surface and gaseous impurities, including water vapour, interfere with these processes and epitaxial relationships are not then observed.

The nature of the layer that covers the metal surface before the nucleation of oxide and which is gradually replaced as lateral growth of the nuclei proceeds is not completely understood. There is evidence[5] that on copper it can be thicker than a monolayer and that it contains metal atoms or ions as well as oxygen atoms or ions. Furthermore, these species have considerable mobility. Lateral growth of nuclei, for example, can occur on thin foils of metal on areas where there is no underlying metal since it has been converted to oxide.

The nuclei must be growing as a result of the surface diffusion on the metal and oxygen species.

If a metal reduces water, the reaction $Me + H_2O \rightleftharpoons MeO + H_2$ has a free energy change that is derived from the difference between chemical potentials of the products and reactants:

$$\Delta G^\circ = \mu^\circ_{MeO} + \mu^\circ_{H_2} - \mu^\circ_{Me} - \mu^\circ_{H_2O} = -RT \ln p_{H_2O}/p_{H_2}. \tag{7}$$

If the initial ratio of partial pressures of these gases is adjusted to those existing under equilibrium conditions, then the mixture will reduce metal oxide (or will not oxidize the metal) since the free energy change of the reaction has been reduced to zero. The necessary ratio p_{H_2O}/p_{H_2} is too small to be attained practically for many oxides.

1.3. OXIDATION KINETICS

(a) Low Temperature

Since the very existence of a metal is determined by the skin of oxide on its surface, and since the reaction of a metal in gaseous media at all temperatures determines the possible uses of that metal, it is clearly important that the mechanism of oxide growth should be understood. The commonest method used for investigating oxidation mechanisms is the determination of isothermal growth-rates of oxides at different temperatures. This is not always easy. Small weight gains (*ca.* 1 μg/cm²) can be difficult to measure accurately. Oxide spalling and significant oxide volatility are among other factors that can impede the determination of useful, accurate and reproducible measurements. The methods and problems of measuring oxide thicknesses are discussed in section 1.8.

Oxidation rates are usually described with reference to the mathematical relationship found between the oxide thickness, y, which is assumed to be uniform, and time, t. At low temperatures and for thin films the relationships are of the form $y = K_1 \log t$, logarithmic, $1/y = K_2 - K_3 \log t$, inverse logarithmic, and $y = K_4[1 - \exp(-K_5 t)]$, asymptotic, where $K_1 - K_5$ are constants. Thicker films which are called scales usually grow according to relationships of the form $y^2 = K_6 t$, parabolic, and $y = K_7 t$, rectilinear, where K_6 and K_7 are constants, and sometimes a combination of the two, paralinear. The difference between a thin film and a scale is often described in terms of oxide thickness, y, and growth rate, although there is no fixed division.

Before considering these relationships and providing explanations of them, the initial combination of oxygen with a bare metal must be considered.

The reaction of a bare metal surface with oxygen is extremely rapid, even at room temperature. Some experiments at very low temperatures have shown that the uptake of oxygen is proportional to the logarithm of the exposure time[6]. The physical and chemical processes occurring during these early stages are complicated and while several hypotheses have been advanced to explain them some common general phenomenological features have been discerned[7]. A very simple explanation is given here.

Initially there will be physical adsorption of oxygen molecules on the bare metal surface. These are relatively loosely bound and the energy of the adsorption process is small, < 25 kJ/mole (6 kcal/mole). These molecules then dissociate and the atoms become much more strongly bound by a process of chemisorption which occurs with a much higher energy charge, > 209 kJ/mole (50 kcal/mole). There is evidence[8] from low energy electron diffraction (LEED) that the chemisorption of oxygen is associated with the movement of a specific number of metal atoms into the plane of the adsorbed oxygen atoms. Together these species form a very stable structure consisting of both negative and positive species which several workers have shown to be more stable thermodynamically than the bulk oxide. Nickel crystals, for example, when heated to near the melting point have been shown[8] to cease exhibiting diffraction patterns of NiO but still show the diffraction patterns of an adsorbed layer. The transition of the monolayer to crystalline oxide must be explained[7] by the effect of a second outer layer of chemisorbed oxygen *molecules* in altering the free energy balance so that the oxide becomes more stable than the monolayer.

Across very thin layers of oxide a very strong electric field is created. Since an oxidation potential is of the order of 1 V, across a film, with a thickness of *ca.* 1 nm (10 Å), there will be a field of *ca.* 10^6 V/nm. Under this field oxide growth can be expected. Metal ions are pulled through the film. As the film thickens, the effect of the field diminishes rapidly and is no longer sufficient for further growth. Unless there is sufficient thermal energy present to cause continued growth by ionic diffusion under a concentration gradient in the film, the rate of oxidation will fall off rapidly. The final thickness formed at room temperature will be 1–4 nm (10–40 Å) and this is the usual range of thickness. After a month or so growth will virtually cease. As the oxidation temperature is raised, the final thickness will be increased, but as long as growth is due to the existence of a high electric field, the oxide will never grow very much. Beyond a certain thickness the growth may be proportional to $t^{1/3}$ or $t^{1/2}$, depending upon the type of oxide, as discussed in section 1.5. Theoretical explanations have been given for this[9].

It is important to observe at this point that logarithmic growth is found in films that are very much thicker than it is possible to form under a high electric field. Explanations are usually based upon physical changes that occur within the oxide. One common example arises where vacancies in a film condense to form cavities, as shown in Fig. 4[10]. Assuming that there is no transport of matter across the cavities, the oxidation paths will be constricted as the film thickens. Usually the growth-rate of this type of film becomes logarithmic only gradually and the initial growth-rate is much faster, being associated with relatively high temperatures. Other explanations are based upon 'ageing' of films, producing changes that are not well understood, but which produce logarithmic growth-rates. Logarithmic growth is usually associated with thin films at low temperatures and all further discussion on this type will be confined to them. Some thin films and scales exhibit a growth that is slower than any form of logarithmic description, and this is referred to as asymptotic. For thin films, difficulties in measurements and the short time of experiment make this type of growth difficult to discern. There is no established mecha-

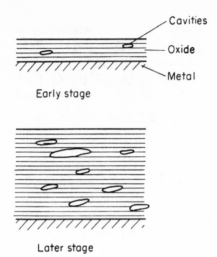

Early stage

Later stage

FIG. 4. The drawing of the elongated cavities found in some oxide films over a period of time. These produce a logarithmic growth-rate[10].

nism for its explanation. Metals at room temperature are often referred to as oxidizing by a logarithmic law, but over many years the weight increase may follow asymptotic behaviour. For thick films, asymptotic growth rates are associated with high residual stresses in the oxide and cavity formation, sometimes causing detachment of the scale from the metal without providing access for the oxidizing gas. The formation of protective oxides on alloys, as described in section 1.6, can also give rise to asymptotic thickening rates, but these arise from the formation during oxidation of an oxide of a different metal.

The actual rate-determining reaction in thin film growth is unknown. There are several possibilities. In oxides with poor conductivities, e.g. Al_2O_3, the passage of electrons through the film is postulated as occurring by the *tunnel effect*, a quantum-mechanical phenomenon which predicts a finite chance that an electron can pass through a thin film of an insulator without requiring a large activity energy. Unusual valencies have been observed in very thin films of other metal oxides. These suggest that surface metal atoms may undergo uncommon ionization processes. Passage of ions through these films, transfer of electrons from metal to oxide, the several steps that occur in the chemisorption process, are other factors to be considered. Since oxide properties vary tremendously, it is unlikely that any one theory will explain this initial oxidation behaviour for all metals. It is possible that each of these factors indicated above may be the most important in different cases.

(b) High Temperature

At higher temperature ranges the rate of oxidation ceases to obey a logarithmic type law. Growth now becomes more rapid and above some temperature follows a parabolic law, generally described as $y^2 = Kt$, where K is the parabolic rate constant and different from any previously mentioned constants. For some metals there is an intermediate temperature range between logarithmic and parabolic oxidation in which $y^n = Kt$, where $n > 2$.

In parabolic oxidation the driving force for the reaction is twofold. Firstly, there is a concentration gradient across the film and secondly, there is an electric potential gradient. These are responsible for diffusion and migration, respectively, across the film. Since the rate for both is inversely proportional to the thickness, the thickening rate of the oxide,

$$\frac{dy}{dt} = \frac{K_{diffusion}}{y} + \frac{K_{migration}}{y}$$

and $K_{diff} + K_{mig} = K$, the observed parabolic rate constant.

Under the conditions that (a) boundary reactions can be ignored, and (b) the field does not act on the moving particles asymmetrically, the energy dissipated is the same in both cases so that it is legitimate to consider parabolic oxidation from a diffusion or an electrical migration aspect.

The concentration gradient in the film produces a difference in chemical potential across the film which in turn results in the necessary free energy change for diffusion to occur.

From the simple version of Fick's diffusion law:

$$\frac{dm}{dt} = DA\frac{dc}{dx} \qquad (3)$$

where dm = mass of matter diffusing across a distance dx in time dt, D = constant, A = area, and dc = concentration difference across the distance dx.

Integrating,

$$m = DA\frac{dc}{dx}t + \text{constant}. \qquad (9)$$

If m = mass of diffusing species in an oxide, ϱ = oxide density, W = molecular weight of the oxide, M = atomic weight of the diffusing species in the oxide and n = number of atoms of that species in a molecule of oxide, then:

$$m = yA\varrho \times \frac{nM}{W}. \qquad (10)$$

Substituting for m in equation (10) from equation (9) and for dx which is equal to y:

$$yA\frac{Mn}{W}\varrho = DA\frac{dc}{y}t.$$

Therefore

$$y^2 = Kt.$$

The parabolic growth of oxides is commonly found in nearly all metals over some temperature range. It is a thermally activated process and the rate constant, K, is equal to $K_0 \exp -(Q/RT)$, where Q = activation energy for the diffusion process, R = gas constant, T = absolute temperature and K_0 is a constant. The change in rate constant for iron[11], which exhibits parabolic oxidation over the temperature range 250–1000 °C, with temperature is shown in Fig. 5.

If only one ion is diffusing through the oxide during its growth, then the value of Q for the oxidation process will be the activation energy for the diffusion of that ion. This value can sometimes be

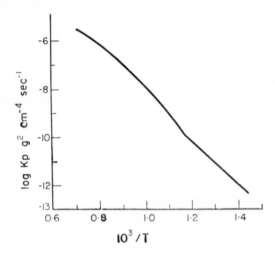

FIG. 5. The variation of the parabolic rate constant for the oxidation of iron with temperature[11].

compared with independently determined values of activation energy for ion diffusion in oxides and thereby provide an important experimental verification for the rate-determining process. For example, above 500 °C the activation energy for the oxidation of copper to cuprous oxide has been determined as 158 kJ/mole (37·7 kcal/mole)[12], which compares well with the separately determined activation energy of 162 kJ/mole (38·8 kcal/mole) for the diffusion of cuprous ion in Cu_2O[13].

In parabolic oxidation the thinner the film the higher is the oxidation rate. Below a certain thickness this is not true and the boundary reactions which are independent of the thickness then become rate-controlling. The early stages of parabolic oxidation are more accurately represented by the equation:

$$\frac{dy}{dt} = \frac{K_1}{K_2 + K_3 y} \tag{11}$$

where K_1, K_2, K_3 are constants. This is called the mixed parabolic equation. At the beginning of parabolic oxidation the thickness is therefore rectilinearly proportional to time, a relationship indicating control by a boundary region, e.g. the rate of movement of metal atoms across the metal/oxide interface. Experimental verification of the mixed parabolic equation is very limited, mainly because it has not been the object of many investigations.

1.4. ELECTRICAL CONDUCTIVITY OF OXIDES

An oxide film grows by ionic transport, but there must also be electron movement going in the same direction as the positively charged metal ions or the reverse direction to the negatively charged oxygen ions. Electron movement through oxides is associated with defects in the oxide lattice as explained below in section 1.5. The ionic diffusion is determined by the movement of ions between vacant lattice sites and is much slower than the movement of electrons. It is largely determined by the concentration gradient, in accordance with equation (8).

The growth of an oxide can be compared with a current flow around a circuit containing an electrolytic cell. It will include both electronic and ionic parts. The ionic current produces two effects:

(a) the ionization of metal atoms, $Me \rightarrow Me^{z+} + z$ electrons, and
(b) the ionization of oxygen atoms, $O_2 + z$ electrons $\rightarrow (z/2)O^{2-}$.

These are respectively anodic and cathodic processes. The metal ions are attracted towards cathodic regions and are called cations, while the oxygen ions are attracted towards anodic regions and are called anions. During oxidation either may be immobile. Oxide growth may occur by cationic movement outwards from the initial metal/oxygen interface or by anionic movement inwards. Some oxides have vacancies in both sublattices and in these both species move during oxide growth. The greatest part of the current is carried by the electrons. These reactions are shown schematically in Fig. 6.

Wagner[14] considered both the diffusion and migration factors that govern the growth-rate and substituted for the former term in electrical equivalents. Hoar and Price[15] considered parabolic growth

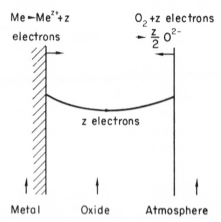

Metal   Oxide   Atmosphere

FIG. 6. The ionic and electronic circuits operating during parabolic growth from which equation (13) is derived[14, 15].

in purely electrical terms: K, the specific conductivity, in Ω^{-1} cm^{-1}, the transport numbers of the anions, cations and electrons, τ_A, τ_C, τ_E respectively, and the free energy decrease of the oxidation reaction, E_0, in volts (from the equation $\Delta G = -zE_0F$). All these quantities can be measured independently.

The total resistance of the circuit, R, is composed of an ionic and electronic resistance combined in series.

$$R = R_{ionic} + R_{electronic}.$$

Therefore

$$R = \left(\frac{\tau_A + \tau_C + \tau_E}{(\tau_A + \tau_C)\,K} + \frac{\tau_A + \tau_C + \tau_E}{\tau_E K} \right) \frac{y}{A}$$

where A = the assumed area of the oxide.

Therefore

$$R = \frac{y}{(\tau_A + \tau_C)\,\tau_E K A}$$

since $\tau_C + \tau_A + \tau_E = 1$.

From Ohm's law, $i = E/R$.

Therefore

$$i = \frac{E_0(\tau_A + \tau_C)\,\tau_E \mathbf{K}A}{y}.$$ (12)

A current i flowing for dt seconds produces $i.dt$ coulombs. These will produce a chemical change, $i.dt.J/\varrho F$ volume of oxide film, where J = equivalent weight of the oxide, ϱ = density of the oxide, and F = Faraday constant, approximately 96,500.

In dt seconds the volume of oxide produced = $dy.A$. Therefore

$$dy = \frac{i.dt.J}{\varrho FA}.$$

Substituting for i from (12) above,

$$\frac{dy}{dt} = \frac{E_0(\tau_A + \tau_C)\,\tau_E \mathbf{K}J}{\varrho Fy}.$$ (13)

From this $dy/dt = K/y$ and $y^2 = Kt$, a parabolic growth-rate. Very close values were found for K by experiment, and K, calculated from equation (13) by substituting into it the values of E_0, τ_A, τ_C, τ_E, and $\mathbf{K}$ independently determined from separate experiments. Some figures are given in Table II[16]. The closeness of the values is a very good verification of the Wagner mechanism, despite the simplifica-

TABLE II. VALUES OF PARABOLIC RATE CONSTANTS DETERMINED (a) FROM EQUATION (13) AND (b) BY DIRECT EXPERIMENT[16]

Compound	°C	Rate constant, equivalent cm^{-1} sec^{-1}	
		(a) calculated	(b) observed
Ag_2S	220	$2\text{–}4 \times 10^{-6}$	$1\cdot6 \times 10^{-6}$
CuI	195	$3\cdot8 \times 10^{-10}$	$3\cdot4 \times 10^{-10}$
AgBr	200	$2\cdot7 \times 10^{-11}$	$3\cdot8 \times 10^{-11}$
Cu_2O	1000	$6\cdot6 \times 10^{-9}$	$6\cdot2 \times 10^{-9}$
($p_{O_2} = 8\cdot3 \times 10^{-2}$ atm)			
Cu_2O	1000	$2\cdot1 \times 10^{-9}$	$2\cdot2 \times 10^{-9}$
($p_{O_2} = 3 \times 10^{-4}$ atm)			

tions of the derivation of equation (13), e.g. E_0 is only constant if there is no polarization, i.e. no 'back emf' created by the 'current' drawn by the oxide, a matter explained more fully in section 2.2. This is only true at the high-temperature end of the parabolic thickening range. It is assumed also that Ohm's law is readily applicable. There will be cases where this may not be true, e.g. for very thin films.

From equation (13) it is clear that the conductivity of an oxide is of major importance and will largely determine its growth-rate under oxidizing conditions. The conductivities of some oxides at 1000 °C are given in Table III[17].

TABLE III. ELECTRICAL CONDUCTIVITIES OF
SOME OXIDES AT 1000 °C[17] (Ω^{-1} cm^{-1})

BeO	Al$_2$O$_3$	SiO$_2$	MgO	NiO
10^{-9}	10^{-7}	10^{-6}	10^{-5}	10^{-2}
Cr$_2$O$_3$	CoO	Cu$_2$O	FeO	
10^{-1}	10^{+1}	10^{+1}	10^{+2}	

As the temperature of a metal is raised, the oxidation rate will increase and be no longer describable by a simple parabolic equation. Some examples need to be quoted to illustrate the several faster rates that are observed.

As an oxide grows, it is constrained by the metal surface on which it forms, since the volume of oxide will not be identical with that of the metal from which it is formed. Stresses are therefore created within the film which may be either compressive or tensile. This was well illustrated by Evans[18] who oxidized a specimen of finely abraded nickel by heating it at one end. The temperature gradient along the strip produced a wedge of oxide which was thickest at the heated end of the specimen. He then removed the oxide by dissolving the nickel underneath and transferred it to a glass plate covered with vaseline, as illustrated in Fig. 7. The thickest part of the oxide wedge wrinkled, showing that it had been under compressive stress when attached to the metal. In thinner regions, e.g. 50–100 nm (500–1000 Å), the film curled, while below 50 nm (500 Å) thickness the film curled up into

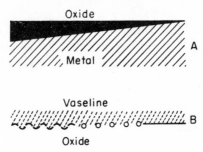

FIG. 7. The appearance of (A) a piece of nickel heated at one end so as to produce a wedge of oxide, and (B) the wedge of oxide after it was stripped from the metal and placed on a glass surface covered with vaseline.

rolls, leaving bare glass surface exposed. The very thinnest parts of the film were not always broken, but this was probably caused by the adhesion of the vaseline.

Much other work has shown that large stresses, both tensile and compressive, are often built up in oxide layers. If these are sufficiently large to cause fracture, then the protectiveness normally imparted to a metal by an oxide film will be lowered. At the point of fracture the residual oxide thickness will be much smaller than the average and the growth-rate will be much higher. If fracture is widespread and repetitive, the overall oxidation rate will no longer be parabolic but will assume an approximately linear rate, as illustrated in Fig. 8, where each break in the curve represents a fracture of the oxide.

The stresses in oxide films are related to the volume ratio of the oxide/metal, Φ, where

Φ = molecular volume of compound MeO/atomic volume of metal Me.

Where this is high, cracking of oxide occurs very readily due to the large difference in volume of the metal and oxide. This is particularly troublesome if the oxygen diffuses through the oxide and forms oxide at the metal/oxide interface where the expansion against the existing oxide will tend to cause detachment.

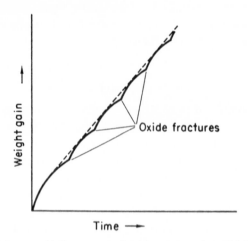

FIG. 8. A linear oxidation rate obtained from a metal on which a parabolically thickening film is continually fracturing.

Both niobium and tantalum oxidize in this way at high temperatures after an initial parabolic growth. The combination of parabolic and rectilinear growth is referred to as paralinear. Other mechanisms also result in paralinear oxidation. Cerium, uranium and tungsten, for example, all appear to form a lower oxide that is oxidized to a higher porous oxide at a constant rate at high temperatures.

Pilling and Bedworth[19] pointed out that if the volume ratio is <1 the oxide formed may not give adequate protection against further oxidation right from the start. Under these conditions, which apply to the light alkali metals over certain temperature ranges, rectilinear oxidation is found, described as $y = kt$.

Under severe conditions, volatilization of the oxide may occur or the heat of activation may be large so that a fast growth-rate results in non-isothermal oxidation and either may give an oxidation curve best described as 'concave upwards'. Extreme cases of oxidation of pyrophoric and highly reactive metals in a finely divided state cause explosions, e.g. titanium. The oxidation behaviour of metals and alloys where localized corrosion occurs leading to sudden increases in weight gain is not always completely reproducible. Sometimes it is referred to as 'breakaway oxidation' and its occurrence indicates that the alloy is unsuitable for use under the conditions employed.

1.5. OXIDE CLASSES

The structures and properties of oxides formed on metals are not always the same as those of the bulk material. A piece of cuprous oxide, for example, has a certain conductivity, lattice structure, chemical constitution, etc., but when cuprous oxide is formed on copper these properties vary through the thickness of the layer. The very existence of a diffusion gradient, which was discussed in section 1.3, indicates this. The amount of oxygen in an oxide is a maximum at the oxide/oxygen interface and decreases to a minimum at the metal. If several oxides form on a metal, the oxide richest in oxygen will be the outermost and the poorest in oxygen will be innermost. This variable composition must be noted, since it is specifically responsible for some of the important oxide properties to be discussed.

Apart from variations in composition arising from concentration gradients, many oxides have an *intrinsic* non-stoichiometric composition. For example, cuprous oxide is generally denoted as Cu_2O, yet chemical analysis indicates that a more accurate formula would be $Cu_{1.8}O$. Charge neutrality is maintained since the oxide contains some cupric ions in the correct proportion. The structure of the oxide must accommodate the cupric ions by the creation of vacant cation sites, since two Cu^+ cations are electrically equivalent to one Cu^{2+} cation. For every Cu^{2+} cation present there must be one former Cu^+ cation site vacant. Since ionic conduction is a diffusion process and is dependent upon the existence of vacant sites, the more cupric ions that are present in the oxide the higher will be its *ionic* conductivity, which will be cationic in nature, i.e. based upon cation movement within the cation sublattice.

The reaction $Cu^{2+} + e \rightleftharpoons Cu^+$ occurs very readily and in cuprous oxide the presence of these two cations permits the easy passage of electrons. This oxide therefore has high *electronic* conductivity too. Since it has already been shown that the oxide conductivity Ω is an important determinant in parabolic growth, cuprous oxide will grow more easily than a more stoichiometric oxide. Equally, cuprous sulphide, Cu_2S, which is more defective than cuprous oxide, being $Cu_{1.6}S$, has a larger rate constant than $Cu_{1.8}O$. This is an important observation. The growth of an oxide depends upon its defective structure. Generally, the electronic transport number τ_E predominates, but at high temperatures the ionic transport number τ_C becomes more important since there is then a considerable amount of thermal activation energy available for diffusion and a larger vacancy concentration.

The conduction mechanism in cuprous oxide arises from an overall electron deficit, is semiconducting in kind (the electrical conductivity increases with rising temperature), and is referred to as *p*-type, i.e. dependent upon positive holes. These should not be confused with the empty cation sites, which are holes in the sense of lattice vacancies, since positive holes refer to empty conduction bands in the outermost Brillouin zone of the oxide. The schematic arrangement of ions in the cuprous oxide structure is drawn in Fig. 9.

Theoretically, at least, it is possible to have *p*-type semiconductor oxides which are not deficient in cations but have an excess number of anions. These would have to be situated on interstitial sites within the oxide lattice and the large size of the anion would produce considerable local strain. No oxide has yet been found that fits into this possible category.

If foreign cations are introduced into the cuprous oxide lattice the conductivity will be altered. If a trivalent cation, e.g. Cr^{3+}, is substituted for a Cu^+ cation the number of cation vacant sites will increase since every Cr^{3+} cation is electrically equivalent to *three* Cu^+ cations or *one* Cu^+ and *one* Cu^{2+} cation. Every Cr^{3+} cation will therefore

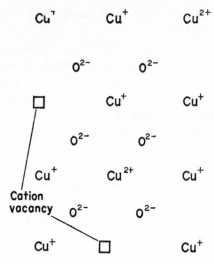

FIG. 9. The arrangement of ions in cuprous oxide.

create one or two vacant sites in order to maintain electrical neutrality. The easy electron exchange referred to above will be adversely affected and the electronic conductivity of the oxide will be lowered. Over a large temperature range electronic conductivity is more important than ionic conductivity so that the overall total conductivity is lowered. At very high temperatures the situation can be reversed. The effect of creating additional cation sites and thereby aiding ionic diffusion predominates and the conductivity of the oxide is raised. Nickel oxide, NiO, is a cation-defective oxide like Cu_2O, containing Ni^{2+} and Ni^{3+} cations. When chromium is alloyed with nickel the oxidation rate increases at 1000 °C since the vacancy concentration of NiO is increased by this addition, as illustrated in Fig. 10. As the proportion of chromium is raised, the oxidation rate reaches a maximum and then starts to fall as shown in Fig. 11. Beyond the proportion of chromium corresponding to the highest oxidation rate, the chromium begins to oxidize to Cr_2O_3 rather than substitute in the NiO lattice. Cr_2O_3 is a highly protective oxide and therefore the overall oxidation rate of the alloys starts to fall beyond this proportion. Eventually the rate

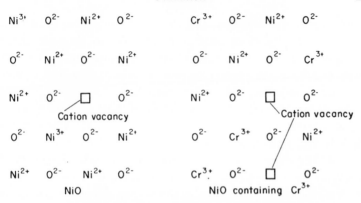

FIG. 10. The arrangement of ions in NiO. The addition of some Cr^{3+} cations raises the cation vacancy concentration of the oxide.

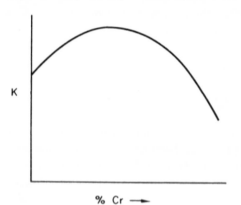

FIG. 11. The variation in the parabolic oxidation rate constant, K, for Ni–Cr alloys with Cr content at high temperature.

is much lower than the original nickel. Nichrome wire (80% Ni–20% Cr), which is used in electric fire element windings, develops a surface which is covered with chromium oxide. This confers very good oxidation resistance at the required dull red heat temperature (800 °C).

If a lower valency cation, e.g. Li^+, is dissolved substitutionally in the NiO lattice, it increases the electron deficit of the oxide since it provides only one electron instead of the two or three electrons provided by the nickel cations. The predominating electronic conductivity is increased since more Ni^{3+} cations must form in order to maintain charge neutrality. Two Li^+ cations are electrically equivalent to one Ni^{2+} cation and the concentration of vacancies in the oxide will be lowered. The ionic conductivity is therefore lowered.

The other main class of oxides consists of those deficient in anions and they therefore contain vacancies in the anion sublattice. An example is Al_2O_3 which has a formula Al_2O_{3-n}, where n is a very small fraction. The cation/anion ratio in the oxide is slightly larger, therefore, than 2/3. Charge neutrality is maintained, since those electrons which are not employed in ionizing the missing anions are maintained within the structure. These extra conducting electrons provide additional conducting bands outside the outermost Brillouin zones. This class of oxides exhibits n-type semiconductivity. Alloying additions affect both the ionic and electronic conductivities, but in a reverse sense from that observed with p-type oxides. Additions of a higher valency cation, Ti^{4+}, raise the electronic conductivity by providing more electrons than the solvent metal ions, thereby increasing the electron/cation ratio. Such additions lower the ionic conductivity, however, since the three tetravalent cations will substitute for four trivalent cations. Rather than introduce cation vacancies into the cation sublattice, such additions lower the cation/anion ratio and in bringing it nearer to the 2/3 ratio in the 'ideal' chemical formula they render the oxide less defective. Additions of a lower valency cation lower the electronic conductivity since the electron/cation ratio is lowered. They raise the ionic conductivity since they occupy more cation sites than the solvent cations and in raising the cation/anion ratio of the oxide further away from the 2/3 ratio they render the oxide more defective.

Some within this class of oxides contain an *excess* of cations, e.g. ZnO, which should accurately be written $Zn_{>1}O$. In this lattice, shown in Fig. 12, there are extra zinc cations, situated in interstitial positions, which supply conducting electrons. Conduction occurs by the movement of electrons from the outermost Brillouin zone to a small number of adjacent unfilled higher energy levels. The presence of excess zinc is

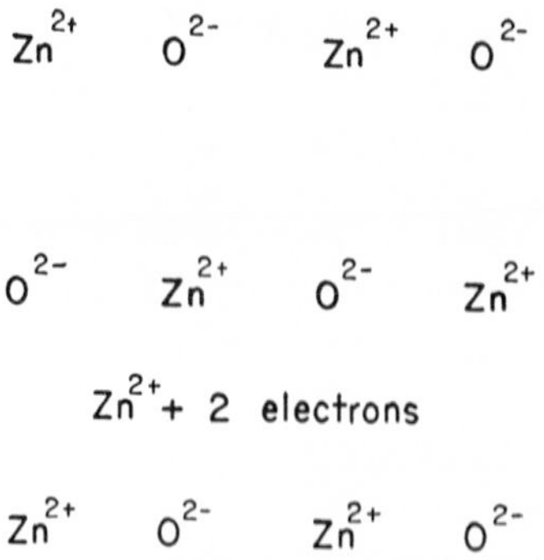

$$Zn^{2+} \quad O^{2-} \quad Zn^{2+} \quad O^{2-}$$

$$O^{2-} \quad Zn^{2+} \quad O^{2-} \quad Zn^{2+}$$

$$Zn^{2+} + 2 \text{ electrons}$$

$$Zn^{2+} \quad O^{2-} \quad Zn^{2+} \quad O^{2-}$$

FIG. 12. The arrangement of ions in ZnO.

indicated during the dissolution of ZnO in dilute HCl by the evolution of small amounts of hydrogen.

As with anion-deficient oxides, if cations of higher valency than zinc are dissolved in ZnO they will contribute extra electrons and therefore raise the conductivity, whereas cations of lower valency than zinc will have the reverse effect.

The validity of these models of oxides, which has been tested by the conductivity measurements described above and performed by Wagner *et al.*[20, 21], has been further substantiated by pressure-dependency effects which have been predicted and observed[22].

Thus for copper, one oxygen molecule adsorbing at the oxide surface must create four Cu^+ cations in the metal which migrate into the oxide, leaving behind four cation vacancies, $4\square$, with four vacant electron sites or positive holes, $4\oplus$, since these electrons have ionized the oxygen molecule.

Thus: $$4\,Cu + O_2 = 4\square + 4\oplus + Cu_2O.$$

From the law of mass action, assuming that the active masses of bulk Cu and Cu_2O are both equal to unity,

$$[\square]^4 \times [\oplus]^4 / p_{O_2} = \text{constant}.$$

Since each vacant cation site is created in association with a vacant electron site, the concentration of cation vacancies, c, is equal to the

concentration of vacant electron sites,
Therefore

$$c \oplus \propto (p_{O_2})^{1/8}.$$

Since the conductivity is proportional to the number of vacant electron sites, the conductivity K should vary according to the eighth root of the oxygen partial pressure. Experimentally it has been found that the conductivity and oxidation rate of copper to cuprous oxide over a certain temperature range are proportional to the $1/n$th root of the partial pressure where $n = 6 \cdot 3$–7, which is in fairly good agreement, once allowance is made for the simplifications involved.

For zinc, oxidation will cause the migration of the existing interstitial cations Zn_i^{2+} and electrons:

$$2\,Zn_i^{2+} + 4e + O_2 = 2\,ZnO.$$

Assuming the active mass of bulk ZnO is unity,

$$[Zn_i^{2+}]^2 \times [\varepsilon]^4 \times p_{O_2} = \text{constant}.$$

Therefore $\qquad c(\varepsilon) \propto (p_{O_2})^{-1/6}.$

Experimental values of the index have varied from $1/4$ to $1/6$.

The value of n has also been obtained[23] from the temperature coefficient of the thermoelectric power vs. $\log p_{O_2}$ by using the equation $dE/dT = -K/en \log p_{O_2} + \text{constant}$, where $K = $ Boltzmann's constant and $e = $ the electronic charge. For Cu_2O, n has been found to be $8 \cdot 5$ (800–1000 °C), and for ZnO, 6 (500–1000 °C), from measurements of the temperature dependence.

At very low pressures other factors dominate the oxidation reaction and the pressure dependency relationship may alter or disappear. If the rate determining factor is, for example, the arrival of the oxygen molecule at the surface and its dissociation, then the ratio will have a different pressure dependency:

Arriving at oxide surface		*Adsorbed*
O_2	$=$	$O^{2-} + O^{2-}$
Gas		(Chemisorbed oxygen atoms)

TABLE IV. THE CHARACTERISTICS OF THE MAIN CLASSES OF OXIDES

Semiconductor	Ionic conduction	Increase in po_2	Presence of higher valency cations in oxide	Presence of lower valency cations in oxide	Example
p-type metal deficit cation vacancies + electron defects	$\tau_c \ll 1$ $\tau_A = 0$	K_e increases	K_e decreases	K_e increases	Cu_2O
n-type metal excess interstitial cations + electrons	$\tau_c \ll 1$ $\tau_A = 0$	K_e decreases	K_e increases	K_e decreases	ZnO
n-type oxygen deficit anion vacancies + electrons	$\tau_c = 0$ $\tau_A \ll 1$	K_e decreases	K_e increases	K_e decreases	Al_2O_3

Applying the law of mass action,

$$\text{Equilibrium constant } K_p = \frac{[O^{2-}]^2}{p_{O_2}}.$$

In the formation of oxides by chemisorption the number of O^{2-} atoms is proportional to $(p_{O_2})^{1/2}$, a dependency found in the oxidation of tantalum at 6–900 °C where p_{O_2} is 1–400 mm. A similar law is found if the gas dissolves in the bare metal after dissociating into two atoms.

For many n-type oxides, TiO_2, ZrO_2, Nb_2O_5, Ta_2O_5, MoO_3, WO_3 and Al_2O_3, the oxidation rate becomes independent of pressure above a certain value since the number of defects is small and they become completely 'saturated' with oxygen at comparatively low pressures.

Many oxides fall into these two main categories based upon p-type or n-type conduction. A few oxides, e.g. CuO, appear to fit into both and have an equal number of interstitial electrons and electron defects. The conductivity of cupric oxide is independent of the partial pressure of oxygen, while dissolving either Cr^{3+} or Li^+ cations in the oxide raises its conductivity. Other oxides fall into neither category but are less important. The characteristics of the main types of oxides are shown in Table IV. In the analyses of doping effects it should be noted that the assumptions are made that the solute cation is uniformly distributed throughout the oxide and that it exhibits its usual valency. When either of these assumptions is untrue, doping characteristics may be different from the expected.

1.6. OXIDATION OF ALLOYS

When a single phase binary alloy consisting of metal A containing an alloying element B is oxidized there are two possibilities. Either (a) a single phase oxide is formed of A containing some B cations or of B containing some A cations, or (b) the two phases of A oxide and B oxide both form. Where only one oxide forms, e.g. AO, the alloy composition at the alloy/oxide interface will change unless A atoms diffuse from the bulk of the alloy to the interface as rapidly as they are removed into the growing oxide. The kinetics of oxide growth and the kinetics of alloy diffusion will determine whether depletion of A occurs, resulting in enrichment of B in the alloy layer adjacent to the interface. If there is gradual enrichment of B, the oxide AO will continue to form alone until the interfacial concentration of B reaches a fractional concentration of n_B^* corresponding to the three-phase equilibrium: $AO + BO + alloy$.

At equilibrium the fractional concentrations of A and B are n_A^* and n_B^*. For AO to form exclusively, the initial concentration of A must be $> n_A^*$. Assuming thermodynamically ideal conditions and ignoring possible nucleation effects, the limiting *mole fraction* of A (N_A) above which only AO will form is given[24] by the equilibrium condition:

$$\Delta G_{AO}^* - \Delta G_{BO}^* = RT \ln \left[\frac{N_A}{1 - N_A} \right] \qquad (14)$$

where ΔG_{AO}^* and ΔG_{BO}^* are the free energies of formation of the oxides AO and BO respectively.

When the interfacial concentration of B reaches n_B^*, BO formation can be expected, although it must be emphasized that such occurrences may be hindered by nucleation kinetics. Under conditions where only A is being oxidized, the concentration of A at the alloy/oxide interface is mainly[24] dependent on (i) the ratio of the oxidation rate constant to the alloy interdiffusion coefficient, (ii) the ratio of the concentration of A in the oxide to that in the bulk alloy, and (iii) the Pilling and Bedworth ratio for the oxide.

Subsequent to the nucleation of BO during the growth of AO, there are many possible reactions. If BO forms a complete and coherent layer, subsequent oxidation will depend upon the permeability of BO by A ions and by oxygen ions. If BO does not form a complete layer, the formation of AO will continue and the scale will consist of AO containing islands of BO. Reaction at the oxide/oxide interfaces may occur, e.g. $Cu_2O + Al_2O_3$ form $CuAlO_2$ and $CuO + Al_2O_3$ form $CuAl_2O_4$[25].

Oxidation resistant alloys are often based upon an alloy in which the solute element has a much greater affinity for oxygen than the solvent element. A typical example is the Cu–Al alloy system with 10 wt % Al. When these binary alloys are oxidized at 800 °C, cuprous oxide forms very rapidly and cuprous cations cross the alloy/oxide interface into the oxide. The aluminium concentration at this interface increases until it forms a layer of protective oxide. This is highly impermeable to the cuprous cations, which can no longer enter the upper cuprous oxide layer. This is then further oxidized to cupric oxide. The rate determining factor for the onset of this protection is

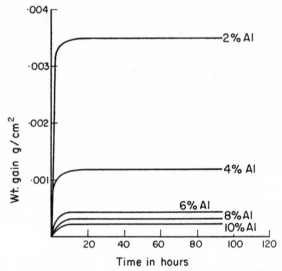

FIG. 13. The oxidation of several Cu–Al alloys at 800 °C[26].

the diffusion of aluminium to the metal/oxide surface where it is oxidized to alumina. The higher the aluminium content of the alloy the more rapidly the oxidation rate (due to the cuprous oxide) is diminished as illustrated in Fig. 13 for a number of binary Cu–Al alloys[26]. Similar behaviour occurs in Cu–Be alloys[27, 28] where a protective layer of BeO forms. The ratio of the proportion of the two copper oxides during oxidation at 500 °C is shown in Fig. 14.

At lower temperatures this protection is not found. In Cu–Al alloys alumina is still formed, but it occurs as particles embedded in Cu_2O

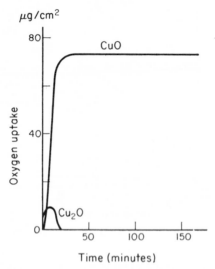

FIG. 14. The variation in the amounts of Cu_2O and CuO during the oxidation of a Cu–Be alloy at 500 °C[27,28].

and not as a continuous protective layer. At the lower temperatures the ratio of the diffusion rates of Cu and Al is lower, but this is not beneficial since the actual value of the rates, which diminish exponentially with the absolute temperature, is a much more significant factor. The aluminium cations cannot arrive at the oxide interface in sufficient bulk to form a complete protective layer and thereby restrict the passage of cuprous cations.

Aluminium also reduces the oxidation rate of iron. This is illustrated in the oxidation curves for several iron–aluminium alloys shown in Fig. 15[29]. The protection is made use of in practice. Since iron–aluminium alloys containing 5–16% aluminium do not have good forming properties, it is not usual to use these alloys directly. Instead, aluminium is deposited on to the surface of the manufactured iron part and diffused into the surface skin by heating. One such process is Calorizing, where an iron surface is heated at 850–950 °C in contact with aluminium powder, alumina and ammonium chloride flux. Other processes include spraying molten aluminium on to the iron surface.

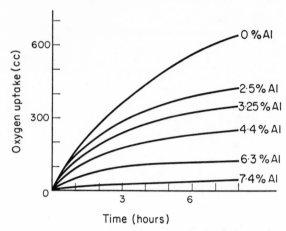

FIG. 15. The oxidation of several Fe–Al alloys at 900 °C[29].

In service, under oxidizing conditions, a highly protective layer of alumina forms on the iron article.

When an alloy containing a solute which has less affinity for oxygen than the solvent metal is oxidized, the solute metal tends to concentrate at the oxide/metal interface. In iron–nickel alloys, for example, a series of iron oxides is formed with nickel cations substituted in them. Any nickel oxide forming either at the interface or even in the scale will be reduced by the iron. The final result is a matrix of iron oxide

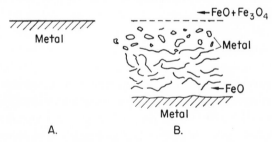

FIG. 16. The appearance of Fe–Ni alloys after oxidation (B). Nickel cations are randomly dissolved in the lattices of the iron oxide, but, in addition, nickel or nickel–iron particles and flakes are found near the metal and within the former metal/atmosphere interface (A).

scale containing nickel cations with small islands of nickel and a metal layer at the interface very rich in nickel. This is illustrated in Fig. 16. The presence of nickel may hinder further oxidation. Since it does not occur in the scale outside the original metal surface, the mechanism is probably one of internal diffusion of anions accompanied by the reduction of any oxidized nickel which varies in form from chains of islands to a fine continuous tracer pattern.

Theoretical considerations[30] indicate that little improvement in oxidation behaviour is to be expected when a metal is alloyed with a very noble solute, e.g. Ni–Pt alloys. Any reduction in oxidation rate is quite small unless the alloying proportion is over 50%. Since the noble element is not oxidized, its concentration at the oxide/metal interface is continually lowered by the diffusion gradient away from the interface.

While the alloying of metals with less noble constituents has been shown to provide the basis for some protective alloys, e.g. Fe–Al, Cu–Al, this is not always true. If the oxide formed on the surface of the alloy is permeable to oxygen, if the oxygen solubility in the alloy is high, and if the diffusion rate of the solute element is relatively slow, then the oxygen may penetrate into the alloy and preferentially oxidize the solute metal. The result will be a layer of oxide particles below the oxide on the alloy surface. This layer is called a subscale and the process internal oxidation. In Cu–0·1% Si alloys it has been found that the thickness of this subscale obeys an approximately parabolic law at 1000 °C[31]. At lower temperatures the oxygen diffuses preferentially down the grain boundaries, which become full of silica. Cu–0·1% Al alloys also exhibit this type of attack. Richer binary alloys in this system form highly protective layers, because there is an adequate mass of aluminium diffusing to the metal/oxide interface. In Cu–Be alloys the same change from protective layer formation to internal oxidation is observed, but the change occurs at a lower proportion of beryllium than in Cu–Al alloys, because the diffusion rate of beryllium is higher than aluminium in copper. In both alloy systems the solute atoms must diffuse to the interface and form a protective layer more rapidly than the oxygen can penetrate into the alloy. In most instances internal oxidation is a nuisance. It alters the surface mechanical properties, and can adversely affect forming operations. Recent developments, however, indicate that it is possible to use the phenomenon deliberately to strengthen metallic lattices.

Double oxides sometimes occur during the oxidation of alloys. These include silicates, which form layers like glass in which diffusion is slow, and spinel phases, consisting of a divalent and trivalent metal (Me' and Me'') in the form $Me'OMe_2''O_3$, which are often formed on ferrous alloys. If spinels are defective in cations, they do not impart any marked improvement to the oxidation resistance of the alloy, but if they are fully stoichiometric then they can have a pronounced beneficial effect. Al–Mg alloys exhibit good oxidation resistance when the oxide formed is $MgOAl_2O_3$ spinel, although the oxidation resistance of aluminium metal itself is also extremely good. Al–Zn alloys are less resistant to oxidation and this is partly because $ZnOAl_2O_3$ is a defective spinel. Spinels usually only form over certain approximate proportions. This is not true for iron where Fe_3O_4 spinel naturally forms. Alloying elements are commonly found in it.

If an alloy contains a solute that volatilizes, then irregular oxidation must occur. When α-brass is oxidized at 700 °C the presence of Zn^{2+} cations in the cuprous oxide should lower the parabolic rate constant, but this effect is short-lived. When the zinc cations penetrate the film, they evaporate from the surface. The final result is a very irregular oxide profile covered with oxidized zinc. This particular loss is minimized by adding aluminium to the alloy. A layer of alumina will form that is highly impermeable to zinc cations, but at higher temperatures Zn cations get through and evaporation returns.

If a binary alloy is oxidized in oxygen at a partial pressure less than the dissociation pressure of one of the constituents, then only the less noble element will oxidize. It does not matter if this element is present in only a small proportion since the oxide of the other is thermodynamically unstable. The diffusion rate of the oxidizable element is also unimportant for the same reason. This special type of oxidation is referred to as selective oxidation and it can cause very remarkable improvements to the oxidation resistance of an alloy.

Price and Thomas[32] oxidized silver alloys containing 0·2–5% beryllium in hydrogen containing 0·1 mm Hg partial pressure of water vapour for 5 min at 600 °C and 20 min at 250 °C. After either of these treatments the surface was covered with a compact film of beryllia and the alloy did not tarnish in sulphur-containing gases which would render the untreated alloy or pure silver black very quickly. The protection afforded was, however, much less than that expected by comparing the conductivities of the natural oxide and pure beryllia. The impurities in the oxide probably raised the conductivity of the latter. The morphology of the selectively formed oxide is important in determining the behaviour of the alloy in tarnishing atmospheres. Cu–Si alloys, for example, exhibited little oxidation resistance by a selective oxidation treatment, after which they oxidized at a rate comparable to that of pure copper[33].

Selective oxidation has the great weakness that the protective film formed is only self-healing in the specific type of atmosphere in which it is formed where the partial pressure of oxygen is too low for the noble element to oxidize. If the film is broken in air, then the normal non-protective film will form and will plug the gap.

1.7. OXIDATION RESISTANCE

In producing an oxidation-resistant alloy, many complex factors have to be considered. Apart from the difficulties involved in providing the resistance itself, there are the practical metallurgical and economic factors to be borne in mind. Any acceptable alloy composition that is arrived at through oxidation experiments must be one that can be easily made and one that can undergo as many forming processes as possible. Expensive alloying elements or manufacturing procedures must be kept to a minimum. Such matters will not be considered further here, but they are very important in any true assessment of practical importance.

In attempting to reduce oxidation rates it is necessary, of course, to investigate at some length the oxidation kinetics of the particular alloy. After that there are two courses that can be adopted. Either the usual film that forms can be modified by a 'doping' process, the object of which is to lower its combined ionic and electronic conductivity, or, the material can be alloyed so that a film of good known protection is produced. Sections 1.4 and 1.6 have described these two approaches.

There is a distinct difference between an alloy where the oxidation rate of the solvent metal is retarded by 'doping' the oxide with solute ions and one where the solute metal forms a separate highly protective oxide layer. In the first case the parabolic rate constant is lowered and, other things being equal, the decrease will be greater as the solute concentration is raised. The limiting factors are the formation of the solute oxide directly, giving a two phase film, and the temperature, since the electronic/ionic conductivities ratio demands different alloying elements of opposite valency, depending upon whether it is greater or smaller than unity. This has already been cited in the case of Ni/Cr alloys. Since NiO is a *p*-type oxide, adding Cr^{3+} should lower its conductivity where electronic conduction is predominant. At high temperatures ionic conduction is predominant and the additional vacancies, created by the presence of Cr^{3+} cations, have an adverse effect upon the oxidation rate constant, as shown in Fig. 11. In the second case, the higher the temperature and the higher the solute content the more quickly the protective solute oxide layer can form. The solute oxide usually has an oxidation rate constant that is several orders of magnitude lower than the solvent metal oxide and it may even be logarithmic in kind. Generally the commercially available oxidation-resistant alloys gain their protection from solute oxides, e.g. Cu–Al, Fe–Cr, Ni–Cr, but the doping of the solvent metal oxide is important too and both aspects must therefore be considered in designing protective alloys. In service, conditions commonly include thermal fluctuation arising from switching equipment on and off. Alloys cannot therefore be chosen merely from their behaviour under isothermal conditions. Some heat-resistant alloys include elements that promote scale adhesion, e.g. additions of yttrium to Fe–Cr alloys.

Stainless Steels

Before discussing stainless steels it is necessary to examine the oxidation of iron.

Under equilibrium conditions iron can form three oxides which are related according to the Fe–O phase diagram[34] shown in Fig. 17.

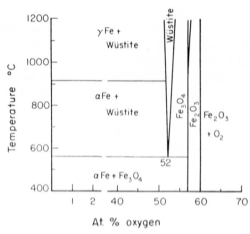

FIG. 17. Part of the Fe–O equilibrium phase diagram[34].

Wüstite, FeO, is a very defective p-type semiconductor which is only stable above 570 °C. It should be noted that even on the equilibrium phase diagram in Fig. 17, the atomic proportions at 570 °C are not 50 : 50 but 48 Fe to 52 O.

Magnetite, Fe_3O_4, is also a p-type semiconductor oxide which has a much lower conductivity than wüstite. It has a spinel structure and is sometimes represented as $FeO.Fe_2O_3$. Both anions and cations diffuse in magnetite.

Haematite, Fe_2O_3, is an n-type oxide with anion defects. It has two structures: α, which has a rhombohedral structure, and γ, which has a spinel structure similar to Fe_3O_4 with which it forms a continuous solid solution.

Between 250° and 1000 °C, iron oxidizes according to a parabolic law and above 700 °C the three oxides have thicknesses in a constant ratio, *ca.* 100 : 5 : 1 FeO : Fe_3O_4 : α-Fe_2O_3[35]. Below 570 °C, where FeO decomposes, oxygen can penetrate the other two oxides and enter the metal.

At low temperatures, oxidation follows a logarithmic law.

If the partial pressure of oxygen is lowered below the dissociation pressure of Fe_3O_4 and α-Fe_2O_3, only FeO will form. Under these conditions, FeO grows according to a linear oxidation rate[36]. This is very unusual, since a dense compact film of FeO is formed, not porous nor cracked like other thick films that grow at linear rates. In this particular case the diffusion rate of cations through the lattice is so large that some other process is rate determining. It has been suggested[36] from other experiments that this process is part of a chemisorption reaction.

Stainless steels, which are among the best commercially available oxidation-resistant alloys, may be ferritic, martensitic or austenitic. They are based upon Fe–Cr alloys and to be fully effective they must contain sufficient chromium to form a layer, consisting largely of Cr_2O_3, which acts as a barrier between the oxygen and the underlying metal. If the proportion of chromium is less than that required to give complete protection, the chromium dissolves in the oxides formed on the iron. Preferential oxidation of chromium is thought to be absent since (a) the reduction of wüstite by chromium is slow and (b) the diffusion of chromium to the interface is slow. At first iron oxidizes and only the innermost oxide, wüstite, contains chromium which forms islands of spinel $FeOCr_2O_3$ in an FeO matrix. At a later stage chromium enters the magnetite spinel lattice to form $FeOFe_{2-\chi}Cr_\chi O_3$ and then the haematite lattice to form $Fe_{2-\chi}Cr_\chi O_3$. As the chromium content in an alloy is increased, the amount of chromium reaching the metal/oxide interface is raised and eventually preferential oxidation to Cr_2O_3 occurs.

The important factor in stainless steels is that the protective oxide is self-healing (unlike, for example, protective films formed by selective oxidation), so that these materials can be worked without losing their oxidation resistance. The proportion of chromium atoms present in the alloy and their diffusion rate must together ensure that there is an adequate number of Cr^{3+} cations to reform the film when it is broken. In aqueous atmospheres the passivity properties of the chromium are also important and these are discussed in section 2.7.

Apart from Cr_2O_3, the film may contain ferrous ions, in the spinel phase $FeOCr_2O_3$. Traces of manganese in the steel become concentrated in the film as a spinel phase $MnOCr_2O_3$ and also as a subscale adjacent to the alloy/oxide interface. The actual composition of protective films on stainless steels varies considerably, depending upon the composition of the alloy, including very small alloying additions, and the oxidation temperature.

The establishment of a fully protective scale takes time and the underlying alloy may be depleted of the protective solute element if this

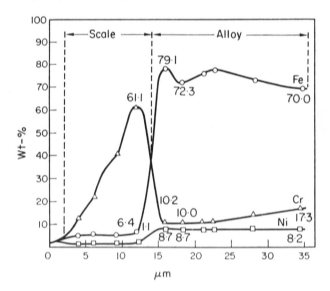

Fig. 18. The distribution of the three major alloying elements within a protective scale and the alloy immediately adjacent to it on an austenitic stainless steel oxidized in air for 18 hr at 1200 °C. Equilibration has not occurred, so that the alloy at the interface is lower in chromium than the bulk of the alloy[37].

element oxidizes more rapidly than it diffuses within the alloy. Fracture of the scale during this period of depletion will result in a higher oxidation rate of the underlying depleted alloy. Since there are a number of processes occurring simultaneously at varying rates, the consequent oxidation behaviour can be very complex.

Typical examples[37] of the varying composition of both scale and adjacent alloy are shown in Figs. 18 and 19, which illustrate element profiles drawn through the protective scale on an 18 Cr–9 Ni austenitic stainless steel, after oxidation for 18 hr at 1200 °C in air. At this temperature the protective scale breaks down and various forms of breakaway oxidation are observed, but the profiles are drawn through part of the protective scale still remaining, which consists mainly of chromium oxide. The profiles are obtained by X-ray probe microanalysis. Since light element analysis is difficult, an oxygen profile is

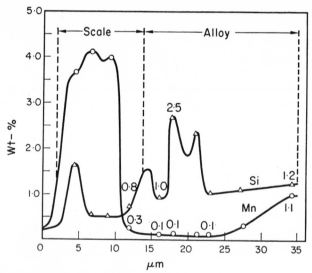

FIG. 19. The distribution of two minor alloying elements within a protective scale and the alloy immediately adjacent to it on an austenitic stainless steel oxidized in air for 18 hr at 1200 °C. Silicon has concentrated in the subscale and towards the outer part of the scale. Manganese is present in only small amounts in the alloy at the interface, but is concentrated in the scale[37].

not shown and the amount of oxygen present can be calculated at any particular point by the difference between the total percentages of elements present at that point and 100%. The nature of the compounds present is not given by this technique, e.g. whether Si exists as silica or silicates, and this must be determined by other means. From the figures, a number of interesting points can be discerned. In Fig. 18 equilibration of chromium has not occurred and the level of chromium in the alloy near the interface is well below the bulk value of the alloy. Failure of the scale at this time would expose a less oxidation-resistant alloy underneath. In Fig. 19 silicon enrichment on the alloy side of the oxide/alloy interface has occurred as a subscale of SiO_2 and in the scale along with manganese, where a complex silicate may have formed and possibly $MnCr_2O_4$ spinel.

The proportion of chromium required to give complete protection depends upon the condition of use. In aqueous environments 12% chromium is required to produce self-passivation, an operation, to be explained later, that produces very thin protective films consisting largely of Cr_2O_3. In gaseous oxidizing conditions, 12% chromium gives excellent resistance below 1000 °C, and 17% above 1000 °C.

Since the resistance is based upon the formation of Cr_2O_3, the initial oxidation rate of the 12% chromium alloy is very high until the protective layer forms. The film is then as thick as 100 nm (1000 Å). With higher chromium proportions this thickness is less.

Aluminium Alloys

Aluminium is covered with a highly protective oxide film of alumina, Al_2O_3. This is an n-type semiconductor containing anion defects, and its conductivity is extremely low. Up to 450 °C the film is amorphous, i.e. it is not possible to detect its crystal structure since the size of the individual crystals is too small, perhaps even unimolecular. Above this temperature the crystallite size grows to 0·5 μ (1 micron, μ = 10^{-4} cm) at 500 °C, then appears to become amorphous again above that temperature and then crystalline again at about 700 °C. The growth-rate of the film at 400 °C is parabolic, but as Fig. 20

shows, the total weight gain is comparatively small[38] and the mechanism is thought to be due to the field strength across the film rather than Wagner diffusion[39]. At higher temperatures the growth is perhaps paralinear and then asymptotic, although the explanations for these changes are not clear. There is no change in oxidation behaviour between the solid and molten metal below and above

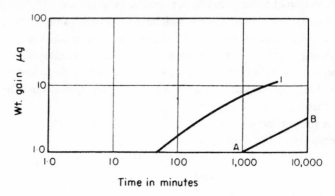

FIG. 20. The weight gain of aluminium at 400 °C/cm² of geometric area[38]. 1, experimental curve. *AB,* parabolic slope.

660 °C, which is the melting point of aluminium. Since the rate controlling process at this temperature is ion movement through the film, and since both solid and molten aluminium are in the same close-packed structural state, no change would be expected.

Aluminium has such good oxidation resistance that alloying is usually more concerned with improving the strength of the metal by age-hardening mechanisms than with lowering the oxidation rate. In this category, Al–Cu alloys are liable to attack about the $CuAl_2$ clusters and precipitates, since the presence of copper lowers the resistance of the film considerably. This is more important in aqueous environments since these alloys are not used much above room temperature in the aged condition. Al–Mg alloys form MgO above 350 °C and are yellow to black in appearance. Between 200–550 °C the oxidation rate is paralinear and similar to that of pure magnesium.

Titanium Alloys

The oxide on titanium, TiO_2, titania, has three forms: rutile, anatase and brookite. Rutile is the natural oxide on titanium and its alloys. This is an *n*-type anion-defective oxide. Although cations with a valency greater than 4 reduce the oxidation rate the effect is small, particularly as those elements, e.g. Mo and W, have less affinity for oxygen than titanium itself. Small amounts of chromium increase the oxidation rate considerably, but with 17% Cr there is a slight improvement. Lower oxides are found at high temperatures.

Growth up to about 500 °C is logarithmic, but since the thicknesses reached are too great to be explained by electric field transport, other explanations, e.g. cavity formation, must be sought, although no mechanisms have been definitely substantiated. Above 550 °C the growth rate is parabolic, with oxygen ions diffusing inwards. Above 850 °C a paralinear growth is found. The scale consists of an inner compact layer of constant thickness and an outer porous thickening scale. Initially, thickening follows a parabolic rate, but after a period of time the rate remains constant, corresponding to the formation of the outer layer.

Copper Alloys

Copper oxidizes according to a logarithmic law up to around 200 °C. Above that temperature there appears to be a cubic law which is operative over a small range and then a parabolic law is found. It was pointed out in section 1.3 that the value of the activation energy, 158 kJ/mole (37·7 kcal/mole), found from the variation of the rate constant above 550 °C, corresponds to the activation energy for the diffusion of cuprous ions in Cu_2O. Below this temperature the value is much less, 83·6 kJ/mole (20 kcal/mole), and is thought to be associated with a reaction in CuO[12].

The proportion Cu_2O/CuO found experimentally is shown in Fig. 21[40]. This is pressure sensitive and the amount of CuO decreases with decreasing oxygen pressure.

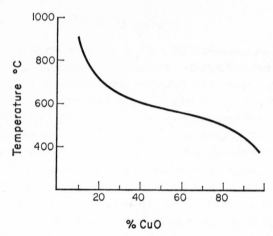

Fig. 21. Approximate variation of Cu_2O/CuO ratio with temperature for the oxidation of copper[39].

Aluminium, beryllium and magnesium increase the oxidation resistance of copper considerably, mainly by preferential oxidation, already described in section 1.6. Many binary alloys, Cu with Ca, Cr, Li, Mn, Si or Ti, oxidize at a similar rate to copper and grow a double scale consisting of outermost CuO and innermost mainly alloy oxide. Copper–zinc alloys have an oxide consisting of Cu_2O matrix with ZnO particles, which form a continuous outermost film when the zinc content is 20%. At low temperatures the presence of zinc cations lowers the growth rate of Cu_2O, while at high temperatures zinc evaporates away when it has permeated the cuprous oxide film. Copper–nickel alloys oxidize at the same rate as copper up to 30% nickel. Many dilute copper alloys suffer from internal oxidation.

1.8. METHODS OF INVESTIGATION

Investigations of oxidation have been largely concerned with the kinetics of film growth and the morphological properties of scales. For kinetic studies, much depends upon the final thickness of oxide. The greater the amount of oxidation, the easier it usually is to make useful measurements, and there are several comparatively easy methods for determining the thickness of scales. The problems of measuring very thin films that are 1–5 nm (10–50 Å) thick are considerable. Before considering the methods that are available, it is worth while to look into the general problem.

In order to determine growth of very thin films, it is of paramount importance to be able to measure the true surface area of the metal surface. This will consist of the actual geometric area multiplied by a roughness factor. For a very carefully electropolished surface, the roughness factor may be very close to 1, but for a coarsely abraded specimen it may be as large as 10. The thickness, if calculated by a measurement involving surface area, can clearly be a long way out unless the roughness factor is known.

1. *Thermal microbalances*

It is possible to make microbalances, weighing to a sensivity of 10^{-7} g, which will continuously measure and record the weight of a specimen in a controlled atmosphere at a carefully maintained temperature. These are made of silica and are often suspended on torsion wires, since these are less likely to vary than a knife edge suspension. Considerable experimental skill is required for their successful operation.

2. *Ellipsometer*

When polarized light is reflected from a metal surface coated with a metal oxide, the plane of polarization is partially rotated. An ellipsometer is an instrument that can measure this rotation which can then be related to the oxide thickness. This is a very delicate technique and although it is complicated it is one of the best methods available. It has been used to measure very small increments, e.g. the room temperature oxidation of tin, which grows from 1·4 nm to 1·9 nm (14 Å to 19 Å) in 1 hr. The instrument can be mounted outside the reaction chamber and oxide growth followed throughout the length of the experiment.

3. *Loss of reflectivity*

The actual loss of reflectivity of a metallic surface can be measured and used for oxidation studies. This technique has not been widely used.

4. *Electrometric reduction*

Some oxides can be cathodically reduced at potentials above that of hydrogen evolution and their thickness can be determined in this way. An example is shown in Fig. 22 for the reduction of oxides on copper. If a constant current is used, preferably recorded, then

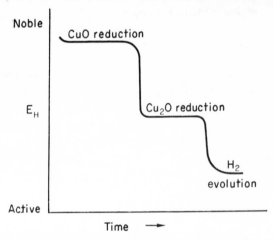

FIG. 22. Idealized potential/time curve for the electrometric reduction of an oxidized copper specimen.

$i \times t$ = number of coulombs required to reduce the oxide. If A = area of film, J = electrochemical equivalent of the film, y = thickness, ϱ = density of the oxide, then $Ay\varrho = it \times J$, from which y can be determined. Since the reduction reaction for oxygen ($O_2 + 2\,H_2O + 4$ electrons $\rightleftharpoons 4\,OH^-$) occurs at potentials above the potential of hydrogen evolution, it is necessary to exclude oxygen from the solution otherwise some of the current measured will be used for this other cathodic reaction. Also, the solution employed must not dissolve the film chemically.

The cathodic reaction for the reduction of Cu_2O is:

$$Cu_2O + H_2O + 2 \text{ electrons} \rightleftharpoons 2 Cu + 2 OH^-.$$

The method has also been employed for determining the thickness of Fe_2O_3 films where the cathodic reaction yields a ferrous ion by a process of reductive dissolution which is considered in detail in section 2.3. The cathodic reaction for this reaction is:

$$Fe_2O_3 + 6 H^+ + 2 \text{ electrons} \rightleftharpoons 2 Fe^{2+} + 3 H_2O.$$

The ferrous ion can then be titrated as a further check upon the thickness calculations. Unlike the other three methods this is a batch process and the construction of an oxidation curve requires many specimens. The accuracy, however, is quite good and the method is widely used. Films as thin as 0·5 nm (5 Å) have been determined in this way.

In thin film work, the initial state of the surface is important. Whether the metal has an air-formed film before starting the experiment will make a lot of difference to the measured thickening rate, although it must be remembered that for many metals this thin film cannot be removed. If the specimen is electropolished before oxidation, then the film left on by the treatment will differ from the film formed in air after etching or abrading. Cold work in the surface will often increase the thickening rate.

For thicker films these factors are less important, but it is in the thin film range that much more information is required to help elucidate the mechanisms of growth.

5. *Interference colours*

When a strip of metal is heated at one end, a wedge of oxide is produced on it as illustrated in Fig. 23. This will cause interference between light reflected from the metal and from the oxide if the path length between these two rays differs by $n\lambda/2$, where n is an odd integer and λ is the wavelength of the light. If y is the thickness of the oxide and μ is its refractive index, $n/2 = 2\mu y$. Usually the metal surface is rough and the oxide thickness varies. Consequently a band of wavelengths is interfered with. As the film thickens it will first interfere with the blue end of the spectrum. The oxide will therefore appear

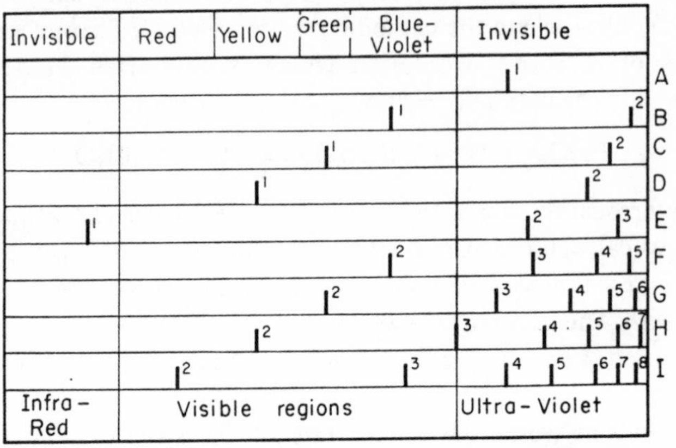

Invisible	Red	Yellow	Green	Blue–Violet	Invisible	
					1	A
			1		2	B
			1		2	C
		1			2	D
	1				2 3	E
			2		3 4 5	F
		2			3 4 5 6	G
		2		3	4 · 5 6	H
	2		3		4 5 6 7 8	I
Infra–Red	Visible regions				Ultra–Violet	

Fig. 23. The production of a sequence of interference colours during an oxidation reaction[41]. The film thickness increases from *A* to *I*. With thickness *A* the interference band (1) occurs in the ultraviolet range and therefore no colour is produced. With thickness *B* the band has reached the blue–violet range and the interference colour is therefore red. As the film becomes thicker (*C* and *D*), the band eliminates light of increasing wavelength. With thickness *E* the first band moves into the infrared invisible region before the second band (2) has moved out of the ultraviolet region. This produces a silvery hiatus between the first order and the second order bands. This gap does not occur again since the bands become progressively closer to each other. The second band and third band (3) are both in the visible range at thickness *I*.

red. Further thickening gives the sequence of colours shown in Fig. 23[41]. Any given wavelength will be interfered with at thicknesses of y in the converging series $1 : 1/3 : 1/5$. It will be seen that the third order interference band enters the visible spectrum before the second has entered the invisible infrared range, and the sequence of colours is therefore altered. Since red and blue are interfered with, the oxide appears green.

If the metal is carefully electropolished before oxidation, the inter-ference band may be so narrow that it does not produce colours. Absorption by the film itself must be very low for this effect.

It is possible to estimate film thicknesses when they cause inter-ference in the visible spectrum. Various factors limit the accuracy of this method, e.g. there is a phase change upon reflection of light from a metal surface, μ may alter with the thickness of the film, etc. Colour comparators have been made, transmitted light through stripped films used, and other refinements employed. Much successful early work was carried out using interference films. In Table V some comparative figures are shown for different thicknesses of iodide films on silver determined by this and other methods[42].

TABLE V. COMPARATIVE FIGURES FOR THE
ESTIMATION BY DIFFERENT METHODS OF IODIDE
FILMS FORMED ON SEPARATE SPECIMENS
OF SILVER[41]

Measured wt. increase (mg/cm²)	Electrometric Calculated wt. increase	Nephelometric Calculated wt. increase
0·162	0·161	0·167
0·106	0·096	0·113
0·059	0·059	0·058
0·018	0·020	0·017
0·002	0·003	0·004

6. Metallographic examination

Many different metallographic methods have been employed in the study of oxidation. Information on the scale about crystallite size, orientation and shape, porosity, cracking, sintering and composition can be obtained with such techniques. One of the main difficulties is the preservation of the oxide while the specimen is being prepared, since oxides are usually brittle and often rather fragile.

The determination of the thickness of oxides and the discrimination of various layers can often be done with an optical microscope. Polarized light can sometimes be used to distinguish different phases. Non-reacting markers help to elucidate ion movement in oxidation processes. In most oxides only one ion diffuses, but in some, e.g. Fe_3O_4, there is definite evidence that both cations and anions move in opposite directions through the film. Other oxides are porous to oxygen. This passes through in the non-ionized state and either reacts with metal underneath or dissolves in it. Markers are often molybdenum or platinum wires. A very early experiment shown in Fig. 24 used Cr_2O_3 on iron[43]. After oxidation the Cr_2O_3 was at the metal/oxide interface, indicating that cations had diffused outwards. Experiments have also been reported with iron wires which finished as hollow tubes of oxide for the same reason.

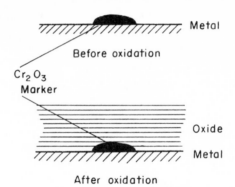

FIG. 24. The determination of ion movement during the oxidation of iron using Cr_2O_3 as a marker[43].

Electron microscope examination of stripped oxide films is often used for studying finer features. The electron probe scanning microanalyser has been used for investigating the quantitative distribution of cations in some oxide films as already indicated in section 1.7. Electron diffraction and X-ray diffraction are widely employed for identification purposes. By combining these techniques together, a very detailed set of information can be obtained. The initial stage of oxidation can be followed by LEED which provides information of the surface structure and by Auger spectroscopy which can be used to identify the elements on the surface.

Many other techniques have been used to a limited extent, e.g. isotope experiments, change in resistance of an oxidizing wire, oxidation under conditions of constantly increasing temperature, measurements of pressure drop in closed reaction vessels. For testing under service conditions, such factors as thermal cycling must be considered, as already indicated, since it is possible that a film that is protective under isothermal conditions may continually spall (break off) under a varying temperature if, for example, the thermal coefficients of expansion of the alloy and oxides are very different.

CHAPTER 2

Aqueous Corrosion

2.1. INTRODUCTION

Aqueous corrosion can take many forms. Apart from general corrosion which results in a relatively uniform removal of a surface, specific features in a metal surface may be preferentially or selectively attacked. Such features include grain boundaries, precipitates and metal/inclusion interfaces. The presence of films on a metal surface may give rise to highly localized regions of corrosion attack, resulting, perhaps, in pitting. Other highly localized forms of corrosion are discussed in Chapter 4. With all these forms of corrosion an anodic and a cathodic reaction must occur, for it was established many years ago that metals corrode in aqueous environments by an electrochemical mechanism. On a piece of metal that is corroding there are both anodic and cathodic sites. These may be permanently separated from each other, but in many instances the whole of the metal surface consists of anodic and cathodic sites which are continually shifting. At an anodic site an oxidation process occurs, which is a loss of electrons, and the metal goes into solution by a reaction that can be depicted as (for example):

$$\text{Fe} \underset{\text{(metal)}}{} -2 \text{ electrons} \rightarrow \underset{\text{(ionized in solution)}}{\text{Fe}^{2+}} \quad .$$

For this reaction to take place a simultaneous reduction process, which is a gain of electrons, must occur at a cathodic site. This reaction consumes those electrons provided by the oxidative process. Unless these electrons can be consumed, then the anodic reaction cannot occur. The reduction of dissolved oxygen and the liberation of hydrogen gas by the reduction of hydrogen ions (particularly from acid solutions) are the two most common reactions occurring during the aqueous corrosion of metals. These can be depicted as:

$$O_2 \quad +4 \text{ electrons} + 2 H_2O \rightarrow \quad 4 OH^-$$
(in solution) (in solution)

$$2 H^+ \quad +2 \text{ electrons} \rightarrow \quad H_2$$
(in solution) (evolved from metal surface)

Other reduction reactions are possible dependent upon (a) the presence of the requisite species, and (b) the electrochemical potential of the metal, as will be discussed later. What is essential is the existence of operative anodic and cathodic reactions both occurring at the same rate, as with any electrochemical cell.

If the metal is partially immersed in a solution, then there is often a permanent separation of the anodic and cathodic areas with the

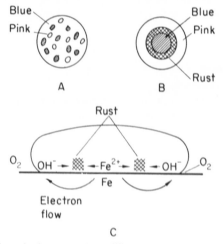

Fig. 25. The salt drop experiment[44]. (A). Initial random distribution of anodic (blue) and cathodic (pink) areas. (B). Final distribution of anodic and cathodic areas with rust formation between them. (C). Reactions occurring within the salt drop.

latter near to the waterline where there is already availability of oxygen. This is illustrated by the salt drop experiment[44].

A drop of 3% NaCl solution was placed on a finely abraded iron surface. The solution also contained small amounts of potassium ferricyanide, which turns blue in the presence of ferrous ions, and of phenolphthalein, which turns pink in the presence of alkali. At first the blue anodic areas and pink cathodic areas formed all over the surface, as shown in Fig. 25 (A). Once all the oxygen in the drop had been reduced, however, further cathodic reaction occurred mainly at the edge of the drop where oxygen from the atmosphere was most readily accessible. The central alkaline regions were not maintained and the whole centre became anodic. The final distribution is shown in Fig. 25 (B). Where the inner anodic and outer cathodic regions met, precipitation of hydrated ferrous oxide occurred which formed an annulus within the drop. Rust formation was initiated by the formation of the $Fe(OH)_2$ which is highly insoluble; it then underwent changes to $FeO.OH$ and possibly other species as discussed in section 2.7.

The reactions within the drop are shown schematically in Fig. 25(C). It must be noted that although the final corrosion product is insoluble, it does not interfere with the dissolution process since it forms away from the anodic sites. If it formed on the metal surface, rusting would quickly stop, since the necessary contact between the metal and solution would be prevented.

If the solution was de-oxygenated before the experiment, e.g. by boiling, then the final distribution shown in Fig. 25 (B) formed at once. The flow of current in the drop could be demonstrated by placing a magnetic field across the drop which then rotated.

It is important to note that neither the anodic nor cathodic reaction in this experiment involves the participation of the sodium chloride. The function of the electrolyte is to provide paths of high conductivity for the ionic current. Other electrolytes could be used and would produce the same effect, provided that they did not react with either the anodic or cathodic products of the corrosion reaction. Magnesium chloride of the same conductivity, for example, would produce the same initial attack, but subsequently there would be the precipitation of magnesium hydroxide on cathodic sites, since this compound has a low solubility product. This precipitate interferes with the cathodic reaction since the electrons causing the reduction of oxygen have to pass through it and the resistance of the circuit is therefore raised. This is an example of cathodic inhibition, dealt with more fully in section 3.4. Since sea water usually contains magnesium ions, it is slightly less aggressive than a sodium chloride solution of the same strength.

This is a very simple picture. The possible effects of insoluble reaction products are many. If the metal surface is vertical or sloping, for example, products will often fall down the surface under the force of gravity. Since they are often associated with local regions of corrosive activity, a matter that is explained later, many metals standing in electrolytes develop vertical corrosion patterns.

A later experiment demonstrated the electrochemical nature of aqueous corrosion quantitatively[45]. A sheet of mild steel was partially immersed in a 3% potassium chloride solution. The anodic and cathodic areas eventually formed as indicated in Fig. 26 (A). Another sheet of steel was then cut along the line that had separated the anode and cathode areas. The two pieces were placed very close to each other but were maintained separate by an insulator and electrically connected

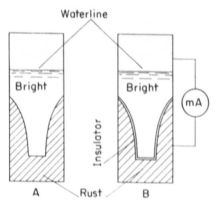

FIG. 26. (A). Distribution of anodic and cathodic areas on a mild steel sheet partially immersed in 3% KCl solution[45]. (B) Simple quantitative determination of current flow in the situation shown in Fig. 26 (A).

through a milliammeter as described in Fig. 26 (B). The current flowing was found to be electrochemically equivalent to the weight loss of the lower anodic portion, which was a very good proof of the electrochemical mechanism of corrosion.

For a specimen of mild steel that is totally immersed, the anodic and cathodic sites are not usually separate. All parts of the surface during corrosion are anodic and cathodic at different times and the whole surface will be covered with a corrosion product which forms away from the metal surface. The electrochemical nature of the corrosion of totally immersed specimens has been shown both by galvanic experiments and by the use of impressed currents. The rate determining factor is often the diffusion of oxygen to cathodic sites. Where this is so, the total weight loss of specimens is then proportional to the area of the liquid surface.

The distribution of corrosion sites and the physical nature of the product both show considerable variation. Sites may cover only a small proportion of a metal, with the remainder of the surface providing a large cathodic area. Under these conditions the sites can be very active, resulting in considerable penetration. This type of attack is called pitting corrosion and is a serious problem on metals and alloys that have highly resistant films, e.g. aluminium and stainless steels, although it can occur on practically any metal surface. The total weight loss of pitted specimens may be very small, but the concentration of attack at a few sites makes it frequently a more serious problem than corrosion attack that may result in greater weight losses but is uniformly distributed over a metal surface.

The physical state of the product of a corrosion reaction can have a large effect upon its progress. Rust forms away from the active sites and has no direct effect. Other products form flocculent or very hard precipitates, depending upon temperature, oxygen concentration, acidity, etc. Pits are often covered with a membrane of product that severely restricts access of oxygen. Within a pit the accumulation of cations will lead to hydroxide precipitation when the solubility product is reached:

$$Al^{3+} + 3\,OH^- = Al(OH)_3 \downarrow.$$

This increases the concentration of H^+ remaining until an equilibrium pH is reached, which will be dependent upon the hydrolysis constants and solubility products of whatever species may be present, e.g. $AlCl_3$. Provided that the volume of solution is rather isolated from the bulk of the solution, considerable changes in cation concentration and pH occur within the pit. The result is increasing acidity and an increasing rate of attack within the pit. In some cases, at least, the mechanism is autocatalytic.

2.2. ANODIC DISSOLUTION

Since practically all metals are covered with an oxide film, this must be removed before the metal can be exposed to an electrolyte. Some oxides are strongly attacked by acids and alkalis, but many are not, and in neutral solutions the removal of oxide films can be rather slow. Some oxides dissolve at a higher rate when in contact with the metal and this important reaction is considered in section 2.8. A metal covered with an oxide has different properties in a solution from a bare metal and the differences can be of critical importance. Nevertheless,

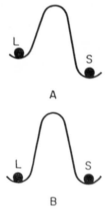

Fig. 27. Adjacent energy wells for a metal cation in lattice (L) and in a solution (S)[46]. (A) The anodic dissolution process predominates. (B) The equilibrium state. Forward and backward reactions occur at the same rate.

in studying aqueous corrosion it is necessary to consider the kinetic processes of metal dissolution and it is simplest to start with the ideal case, a pure, bare metal electrode on which the only reaction occurring is metal dissolution.

The metal atoms on the surface are in energy wells associated with the lattice structure and in order to pass into the solution they have to acquire a certain amount of energy, which is called activation energy. The function of this is to aid the complex process by which the atom is detached from the metal lattice, penetrates the layer of water molecules in contact with the metal surface and acquires a cage of these so that the final product of these reactions is a hydrated metal cation. The initial and final energy wells are represented in Fig. 27 (A)[40], where the latter is lower so that the process proceeds from left to right and the metal dissolves. When equilibrium is achieved, the energy wells will be at the same level, as represented in Fig. 27 (B), and the forward and back reactions correspond to:

$$\underset{\text{(in lattice)}}{\text{Me}} \rightleftharpoons \underset{\text{(hydrated, in solution)}}{\text{Me}^{z+}} + z \text{ electrons.}$$

Anodic dissolution and cathodic deposition are then proceeding at the same rate, and this is described as the exchange current density. For many metals the value is 10^{-1} to 10^{-5} A/cm^{-2}, but transition metals have much lower values.

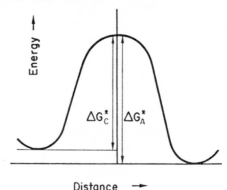

Fig. 28. Energy/distance relationships at a metal/solution interface under equilibrium conditions[46].

Metal cations are more readily dissolved from surface irregularities than from a flat face, since the number of bonds on an atom at such an irregularity are fewer and the packing is more irregular. Dissolution proceeds, therefore, from grain boundaries, emergent dislocations, surface half-planes, etc.

The two energy wells of the initial and final stages are redrawn schematically in Fig. 28 to show the energy/distance relationship of a reversible process[46]. If ΔG_A^* and ΔG_C^* are the standard free energy changes in joules/mole for the formation of the transition state from the lattice and solution sides respectively, corresponding to anodic dissolution and cathodic deposition, and if i_0 is the exchange current density, in A/cm^{-2}, then:

$$i_0 = L_A \exp\left[\frac{-\Delta G_A^*}{RT}\right] = L_C \exp\left[\frac{-\Delta G_C^*}{RT}\right]$$

where L_A, L_C are parameters containing dimensional constants, etc.

If a potential is now applied to the metal so that it becomes an anode, the equilibrium depicted in Fig. 28 will be upset and the situation will correspond to that shown in Fig. 29[46]. The value of the

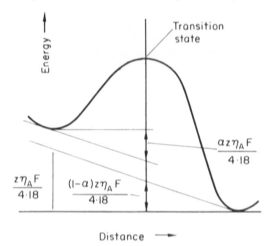

FIG. 29. Energy/distance relationships at a metal/solution interface subjected to an anodic overpotential A[46].

anodic potential is written η_A and is referred to as the anodic over-potential. $\alpha =$ the ratio of the distance between the transition state and the lattice energy well/total distance between lattice and solution energy wells. It is assumed that the alteration is regular over the total distance and that the full effect of the anodic overpotential is made over that distance.

The tilt in the free energy diagram is divided between the anodic and cathodic reactions in the ratio $\alpha/1-\alpha$. The total change in free energy resulting from the applied overpotential is derived from the general equation $\Delta G = -zE_0F$ and in this instance is equal to $z\eta_A F$ joules and therefore $z\eta_A F/4{\cdot}18$ calories.

If i_A is the applied anodic current density in A/cm^{-2}, then:

$$i_A = L_A \exp\left(-\left[\frac{\Delta G_A^* - \dfrac{\alpha z\eta_A F}{4{\cdot}18}}{RT}\right]\right)$$

$$-L_C \exp\left(-\left[\frac{\Delta G_A^* + \dfrac{(1-\alpha)z\eta_A F}{4{\cdot}18}}{RT}\right]\right).$$

Therefore $\quad i_A = i_0\left(\exp\left[\dfrac{\alpha z\eta_A F}{4{\cdot}18\,RT}\right] - \exp\left[\dfrac{-(1-\alpha)z\eta_A F}{4{\cdot}18\,RT}\right]\right).$ (15)

This equation shows that the current drawn by a dissolving electrode arises from the difference between the rates of the forward dissolution reaction and the backward deposition reaction, which becomes neg-ligible as η_A is increased. Depending upon the values of α, z and η_A, the second term will become negligible as i_A increases. If $\alpha = 0.5$, $z = 2$ and $\eta_A = 25$ mV, then at ambient temperatures

$$i_A = i_0 \exp\left[\frac{\alpha z\eta_A F}{4{\cdot}18RT}\right].$$

Therefore $\quad \eta_A = \dfrac{2{\cdot}303\times4{\cdot}18RT}{\alpha zF}(\log i_A - \log i_0).$

Therefore $\quad \eta_A = b_A \log i_A + a_A$ (16)

where $a_A = -b_A \log i_0$.

This is the Tafel equation and b_A is the Tafel constant for the anodic reaction. There will also be a Tafel equation for the cathodic deposition reaction relating the change in potential in the cathodic direction, η_c, and the net current, i_c, between the increasing cathodic reaction and the decreasing anodic reaction which becomes negligible depending upon the values of α, z and η_c.

In order to produce a change in the electrode reaction, so that the current in one direction is greater than that in the other direction, the potential of the electrode must change from the equilibrium value. The change is referred to as polarization and the relationship between the change and the value of the net current density resulting from the change constitutes a polarization curve. Polarization can arise for several reasons, as described below, but the Tafel equation arises from the energy required to change a species from one state to another at an electrode surface at a specific rate. η_A in equation (16) is referred to as activation polarization and *all* electrode reactions involving oxidation or reduction will exhibit it.

If the polarization change is caused by application of an external current then a polarization curve can be plotted as η_A vs. applied

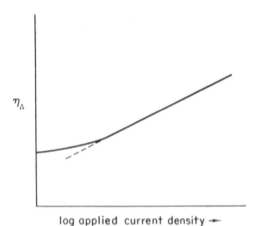

log applied current density →

Fig. 30. An anodic polarization curve exhibiting only activation polarization. The Tafel equation only holds when $i_a \gg i_0$.

current density and a typical plot is shown in Fig. 30. Until $\eta_A > 0.025$, the line is curved since it represents an increasing difference between the two rates of dissolution and deposition as indicated in equation (15). In the case of anodic polarization, the term for deposition becomes negligible when $\eta_A > 0.025$. What is measured is E_i, the potential of the metal electrode at the applied current density i, and i, where $i = i_A - i_C$, the difference between the anodic dissolution and cathodic deposition processes. Consequently $\eta_A = E_i - E_0$, where E_0 is the reversible potential.

The Measurement of Electrode Potentials

A metal establishes a potential with respect to a solution which cannot be measured in absolute terms. The potential *difference* between it and another electrode can be measured. Furthermore, changes in potential difference can be related to the metal electrode under investigation if the other electrode does not change, i.e. is a reference electrode. There are several reference electrodes which are constant provided that the current drawn from them is extremely small, e.g. 10^{-12} A/cm². They are used, therefore, with millivoltmeters of high impedance.

A metal in contact with a solution containing the metal ions at unit activity establishes a fixed potential difference with respect to every other metal in the same condition. The full list of these comprises the electrochemical series of normal electrode potentials of metals in which all the potentials listed are actually differences between the metal and hydrogen, the potential of which is arbitrarily designated as zero. The series of normal electrode potentials at 25 °C is shown in Table VI together with the fixed potentials of some common reference electrodes. The sign convention adopted by the International Union of Pure and Applied Chemistry is used, but it must be pointed out that many workers in America use an opposite sign convention. The reasons for two sign conventions are beyond the scope of this book, but there is often some confusion over the significance of the potential differences that are quoted and over the reality of positive

TABLE VI. STANDARD ELECTRODE
POTENTIALS AT 25 °C

Equilibrium	Normal hydrogen scale (volts)	
Au $\rightleftharpoons$ Au^{3+} +3e	+1·5	
Ag $\rightleftharpoons$ Ag^{+} +e	+0·7991	
Cu $\rightleftharpoons$ Cu^{2+} +2e	+0·337	
H$_2$ $\rightleftharpoons$ 2H^{+} +2e	0·000	
Pb $\rightleftharpoons$ Pb^{2+} +2e	−0·126	
Sn $\rightleftharpoons$ Sn^{2+} +2e	−0·136	
Ni $\rightleftharpoons$ Ni^{2+} +2e	−0·250	
Cd $\rightleftharpoons$ Cd^{2+} +2e	−0·403	
Fe $\rightleftharpoons$ Fe^{2+} +2e	−0·440	
Zn $\rightleftharpoons$ Zn^{2+} +2e	−0·763	
Al $\rightleftharpoons$ Al^{3+} +3e	−1·66	(calculated)
Mg $\rightleftharpoons$ Mg^{2+} +2e	−2·37	(calculated)

Source: Wendell M. Latimer, *The Oxidation States of the Elements and Their Potentials in Aqueous Solutions*, 2nd edition. © 1952 by permission of Prentice-Hall, Inc., Englewood Cliffs, New Jersey.

Reference Electrodes

Ag/solid AgCl in N/10 KCl	+0·288 volt
Cu/saturated CuSO$_4$	+0·316
Hg/solid Hg$_2$Cl$_2$ in N/10 KCl	+0·334

and negative in relation to anodes and cathodes. Some further consideration is necessary therefore.

The free energy of a reaction, ΔG, is related to the potential of a reaction, E_0, by the equation:

$$\Delta G = -zE_0 F.$$

For a reaction to occur spontaneously ΔG is by convention negative and E_0 will have therefore a positive sign. Table VI indicates the relative tendencies of species to ionize, i.e. to lose electrons. Active metals readily become cations in solution. Noble metals become cations less readily by comparison. At the active end of the table, metals provide anodes as a result of the attraction of electrons to the more noble electrode which exhibits cathodic behaviour. Zinc, for example, is anodic to copper. When zinc is put in contact with a solution containing copper, the copper is deposited ('plates out') and the zinc goes into solution in accordance with the reaction:

$$Zn + Cu^{2+} \rightarrow Zn^{2+} + Cu. \tag{17}$$

With the sign convention employed in Table VI the electrons are attracted to the more positive electrode, viz. copper, which is therefore the cathode. The measured potential difference is positive, corresponding in the Daniell cell, which employs these two electrode reactions, to $+0.337 - (-0.763) = 1.1$ V. The two reactions are:

$$(i) \ Zn \rightarrow Zn^{2+} + 2 \text{ electrons}$$

and

$$(ii) \ Cu^{2+} + 2 \text{ electrons} \rightarrow Cu.$$

The electrode potentials of many metals have been measured directly or calculated from thermodynamic data and compiled to give the standard potentials of Table VI.

The table is very useful as a help to understanding many corrosion phenomena, but it is no more than a guide to possible occurrences. In most corrosion reactions, the values shown in Table VI do not apply directly for two reasons. Firstly, metal surfaces are often covered with a film and this will alter the potential of the metal. Secondly, the activity of the metal ions in solution will usually be much less than unity and this too will alter the potential of the metal, as is explained in the next section.

The term activity refers to the effective or thermodynamic concentration. In any electrolyte, the oppositely charged ions exert attraction effects which alter the behaviour of the electrolyte that would be expected from the measured concentration. The more concentrated the solution, the greater is the mutual attraction effect. The correction

factor, γ, or activity coefficient, is used to relate activity, α, with mola-
lity, m, in $\alpha = \gamma m$. As the dilution increases, $\gamma \rightarrow 1$. For many practical purposes activity and concentration are taken as equal, since the
error is small unless the electrolyte concentration is high ($> 10^{-3}m$
very approximately).

In most corrosion reactions, the metal is covered with a film and
frequently the activity of the metal ions is less than unity, so that the
values shown in Table VI do not apply directly.

A simple circuit for drawing a polarization curve is shown in Fig. 31.
The variable resistance R is used to alter the externally applied current.

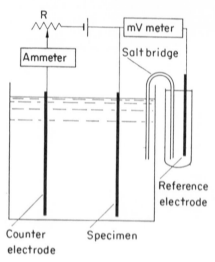

Fig. 31. A simple circuit for drawing a polarization curve.

Usually a short time is allowed to lapse before reading the ammeter
to allow equilibrium to be established at the electrode surface. In some
cases the current density never becomes constant and it is necessary
to make a reading after a chosen time, which should be quoted with
the resulting graph, since another worker making readings after a
different time lapse might obtain different values. The importance of
this point depends upon the nature of the experiment.

The battery shown in Fig. 31 is connected in the circuit so that the metal is an anode. The circuit will therefore produce an anodic polarization curve. Reversal of the battery connections would be necessary to produce a cathodic polarization curve. The auxiliary or counter electrode for making the circuit through the cell should not polarize excessively when current is passed through it, nor should it contaminate the solution. Platinum is therefore often used as the material for this electrode. The reference electrode is frequently silver/silver chloride or mercury/calomel in chloride solutions. To avoid contamination by chloride ions, the reference electrodes are kept outside the cell containing the electrode that is under investigation and connected to it by a salt bridge. This may be agar-agar or wet string, etc., saturated

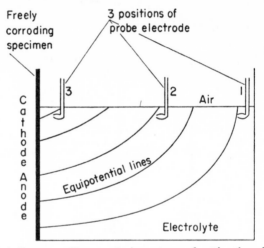

Fig. 32. A diagram to illustrate the importance of putting the reference probe electrode in the correct position. Positions 1, 2 and 3 will give different values.

in an electrolyte whose anions and cations carry approximately equal amounts of current, e.g. ammonium nitrate. This arrangement almost eliminates inconvenient junction potentials. Experimental arrangements often include a Luggin capillary placed very close to the specimen, which permits the measurement of potential at specific sites on an electrode surface. The arrangement is called a probe electrode. Its use is necessary (a) if there is much variation in potential over the surface, and (b) if there is a large polarizing current, since the potential drop between the surface and capillary, equal to current times resistance, will be a function of distance. A simple case is illustrated in Fig. 32, which shows the importance of correct positioning of probe electrodes. Where the electrode surface has a complicated shape and an irregular distribution of current density, this problem can be very difficult.

The Nernst Equation

The values of potential in Table VI refer to metals in contact with solutions containing metal cations at unit activity. It is important to know how these values vary when the activity is different from unity. The relationship is described in the Nernst equation:

$$E_{metal} = E_0 + \frac{RT}{zF} \ln a_{Me^{z+}} \tag{18}$$

where E_{metal} is the potential of the metal, E_0 the standard electrode potential of the metal, R the gas constant, T the absolute temperature, z the valency, F the Faraday constant, and $a_{Me^{z+}}$ the activity of the metal ions.

At 25 °C for a univalent metal, the potential will change 59·9 mV for every tenfold change in activity.

The Nernst equation is derived from thermodynamics[48] and the general form is:

$$E = E_0 + \frac{RT}{zF} \ln \frac{a_{\text{oxidized species}}}{a_{\text{reduced species}}} ,$$

For the single electrode reaction of a metal dissolving this is reduced to that of equation (18) since the activity of the metal is taken as unity. Thus:

$$\text{Me} \rightarrow \text{Me}^{z+} + z \text{ electrons.}$$
$$\text{(reduced)} \quad \text{(oxidized)}$$

As $a_{Me^{z+}}$ is lowered, the potential of the anodic reaction becomes more active. If as a result of dissolution the activity increases, i.e. the dissolution rate creates metal cations more rapidly than they can diffuse away or dissolution is occurring in a restricted volume of solution, the potential of the metal electrode will become more noble. In situations where there is an uneven flow of an environment past the surface, perhaps even a region of stagnation, potential differences may arise between parts of a metal surface in contact with liquid of different metal activities.

For the single electrode reaction of a cathodic process, e.g. oxygen reduction, the Nernst equation is reduced to:

$$E = E_0 + \frac{RT}{zF} \ln \frac{1}{a_{OH-}} \tag{19}$$

where E_0 is that of the oxygen electrode ($+401$ mV E_H) and not the same therefore as in equation (18) and where the activity (fugacity) of the gas is taken as unity.

Thus:

$$\underset{\text{(oxidized)}}{O_2} + 2\,H_2O + 4\,\text{electrons} \rightarrow \underset{\text{(reduced)}}{4\,OH^-}.$$

As a_{OH-} is reduced (corresponding to a falling pH), the potential of the cathodic reaction will become more noble.

It has already been observed that corroding metals are frequently covered with films which may change the potential of the surface. The establishment of such films is a subject of considerable importance, since they can be expected to retard the dissolution rate and thereby impart some degree of protection to the underlying metal. Such protection is commonly described by the term 'passivation' arising from observations on iron by Faraday. When a metal reacts with water it may either form soluble cations or form directly a passive film, and clearly the reaction that occurs is important since the first will result in corrosion while the second will result in less, possibly very much less, corrosion, depending upon the properties of the film.

The thermodynamic data on reactions between many metals and water have been measured, collected and calculated[49] by Pourbaix[47]. They have been combined with solubility data on oxides and hydroxides and with equilibrium constants for reactions of these to produce Pourbaix[47] diagrams which indicate thermodynamically stable phases as a function of electrode potential and pH.

In deriving such a diagram it is necessary to consider four different types of reactions that may occur:

(1) Neither H^+ ions nor electrons are involved, e.g.
CO_2 (aqueous)$+H_2O \rightarrow H_2CO_3$ (aqueous).

(2) Only H^+ ions are involved, e.g.

$$Fe^{2+}+H_2O \rightarrow Fe(OH)_2+2 H^+.$$

(3) Only electrons are involved, e.g.

$$Fe^{2+} \rightarrow Fe^{3+}+e.$$

(4) Both H^+ ions and electrons are involved, e.g.

$$Fe^{2+}+3 H_2O \rightarrow Fe(OH)_3+3 H^++e.$$

Reaction (2) will be pH-dependent, reaction (3) potential-dependent, while reaction (4) will be dependent upon both these variables.

In order to display graphically the reactions of a metal over the whole range of pH, at different values of potential, a large amount of data must be collected. Iron, for example, dissolves to form Fe^{2+} which at higher electrode potentials is oxidized to Fe^{3+} under strongly acidic conditions. The potential for the dissolution reaction will depend upon the activity of the ferrous ions in accordance with the Nernst equation. From Table VI $E = -0.44$ V when the activity $= 10°$. At 10^{-2} $E = -0.44-(2 \times 30) = -0.5$ V. Thus this reaction is represented in Fig. 33[47] as a series of horizontal lines (i.e. different potentials) corresponding to different concentrations of the ferrous ions. The reaction $Fe^{3+}+H_2O \rightarrow FeOH^{2+}+H^+$ is only pH-dependent and from the equilibrium constant of this reaction, the variation of pH with Fe^{3+} ion concentration can be found. In Fig. 33 it is represented by a series of vertical lines (i.e. different pH). The reaction $Fe^{2+} + H_2O \rightarrow FeOH^{2+} + H^+ + e$ is both pH- and potential-dependent and the locus of points at which the $FeOH^{2+}$ and Fe^{2+} activities are equal is the line E (volts) $= 0.877 - 0.0591$ pH[49] and the slope is -59.1 mV/pH unit.

Other reactions involving solid products require a knowledge of solubility product and the reactions of metal oxide and hydroxides under alkaline conditions.

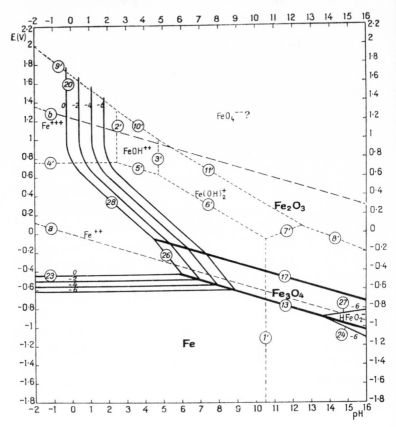

Fig. 33. The Pourbaix diagram for iron[47].

Since the hydrogen and oxygen evolution reactions are both pH-dependent and involve electrons, they also have sloping lines which are described in Fig. 33 by a and b respectively. Between the lines water is stable. Above the oxygen line (b) oxygen is evolved, while below the hydrogen line (a) hydrogen is evolved. In practice it will usually be necessary to take the potential somewhat above line (b) in order to liberate oxygen and somewhat below (a) in order to liberate hydrogen. This additional potential change or overpotential varies from one metal to another and from one environment to another for a given metal, being determined by the properties of surface films and of the metal surface.

The difference in the potential movement required to produce hydrogen and oxygen arises from the opposite reaction involved. Hydrogen evolution includes a capture of electrons, while oxygen evolution includes a loss of electrons:

$$H^+ + \text{electrons} \rightarrow H_2 \text{ (evolved)}$$

$$OH^- \qquad\qquad \rightarrow O_2 \text{ (evolved)} + \text{electrons.}$$

In regions on a diagram where a solid compound is formed, the metal surface may be protected from further attack. This region is then passivated.

A full description of Fig. 33 is beyond the scope of this book, so the three different regions of the diagram, corrosion, immunity and passivation, are represented on a simpler diagram for iron in Fig. 34[47].

With all the reactions that include a pH change, it must be appreciated that as the reactions proceed the local pH will alter, unless the solution is buffered. At cathodic areas, where oxygen is reduced to hydroxyl ions, the pH will rise, an occurrence already noted from the salt drop experiment. Where the metal forms an oxyanion, an additional region of corrosion is found at very high pH corresponding to reactions of the type:

$$Zn + 2OH^- \rightarrow ZnO_2^{2-}.$$

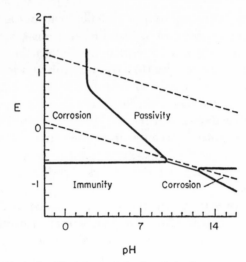

FIG. 34. A simplified form of the Pourbaix diagram for iron[47].

At anodes the pH will start to fall. On aluminium, for example, in neutral solutions, the formation of Al_2O_3 produces anodic acidity which increases the local corrosion rate. Reformation of a protective film is not possible in the acid solution and very intensive attack on localized sites occurs. In this attack the local acidity may lower the potential below that of hydrogen evolution. On iron, the reaction $Fe \rightarrow Fe^{2+}$ starts at a pH just less than 7 and the overall attack is more widespread than on aluminium where the oxide is thermodynamically stable to a much lower pH.

Simplified Pourbaix diagrams for zinc and chromium are illustrated in Figs. 35A, 35B[47]. Zinc corrodes over a wide range of pH, since it forms cations up to pH 9 and oxyanions under alkaline conditions, as already mentioned. A solid film is stable on chromium over a much wider range of pH, but not at noble potentials, since soluble hexavalent chromium ions are formed and corrosion occurs over the whole pH range.

Pourbaix diagrams provide a strong thermodynamic basis for understanding corrosion reactions. It cannot be emphasized too strongly,

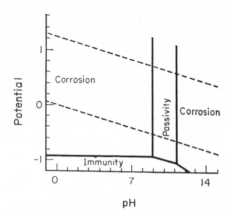

FIG. 35A. Simplified Pourbaix diagram for zinc[47].

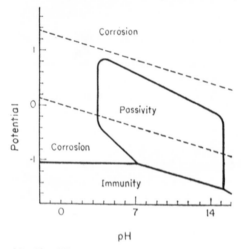

FIG. 35B. Simplified Pourbaix diagram for chromium[47].

however, that they are only a guide to understanding of behaviour and must not be blindly used for prediction. There are two principal limitations on the use of the diagrams. The first limitation arises from lack of kinetic data. No considerations of the kinetic phenomena are included in them, since they are based upon thermodynamic data. In practice it is found that some reactions continue in regions of a

Pourbaix diagram where they are not shown to occur. Sometimes these metastable regions are of critical importance. The regions that are marked Corrosion, Immunity and Passivity represent conditions where under equilibrium conditions corrosion 'can, can't and doesn't' occur, respectively. Passivity can be destroyed but not immunity. The second limitation arises from the purity of the environment. The diagrams are derived from known reactions between metals and pure water. In practice, corrosion problems arise from water containing dissolved salts and the additional reactions that can occur from their presence must be considered also and shown on the diagrams. Successful attempts have been made to do this, e.g. for Brussels tap water in contact with copper[50]. For most situations, however, no such diagrams exist.

Concentration and Resistance Polarization

The Tafel equation is derived purely by considering the activation processes that govern the dissolution of a pure metal. If careful experimental conditions are maintained, the Tafel constant, b, can be found from the straight line position of the experimental polarization curve and the exchange current density, i_0, can be found by extrapolating the straight line back to $\eta_A = 0$, which corresponds to the standard electrode potential of the metal. If experimental values are used to determine Tafel constants and associated information, it is clearly necessary to ensure that no other factors that might affect the results are present. This is not easily done. There are two main factors that hinder the experimental determination of a Tafel slope—concentration and resistance polarization. These must now be considered.

In the simple activation dissolution processes described above, it was assumed that when a cation is dissolved it readily moves away from the metal–solution interface. This movement is a diffusion process and as the dissolution current density is increased the rate of removal of the anodic products does not increase proportionally, so that the concentration of these products in the anolyte (that part of the electrolyte in contact with the anode) will increase and exert a back emf. Further increases in anodic dissolution rate therefore become more difficult and require a disproportionate rise in anodic overpotential. At this stage the rate of dissolution is iA/zF for an electrode of area A and this will be equal to $(AD[C-C_0])/\delta$ where D is the diffusion coefficient, C and C_0 are the ionic concentrations at the electrode surface and in the bulk of the solution respectively and δ is the diffusion layer thickness. From the Nernst equation, the over-potential or polarization arising from concentration effects, η_C, can be calculated:

$$\eta_C = \frac{RT}{zF} \ln \frac{C}{C_0}.$$

Since

$$C_0 = C - \frac{i\delta}{DzF},$$

$$\eta_C = \frac{RT}{zF} \ln \frac{1}{\left(1 - \dfrac{i\delta}{CDzF}\right)}.$$

As $\eta_C \to \infty$ the critical limiting or diffusion current density is given by $(CDzF)/\delta$. Eventually the limiting current density for diffusion is reached. If the anodic overpotential is raised further there will be no increase in current density. The potential change resulting from concentration polarization can be expressed as:

$$\eta_{(concentration)} = 2\cdot3 \frac{RT}{zF} \log \frac{i_L - i_A}{i_L}$$

where R, T, z and F have the usual meanings, i_L = limiting diffusion current density for the forward reaction and i_A = externally applied current density. As the potential is made more noble, any further increases in current density will occur only if other reactions are initiated. This is often oxygen evolution.

The value of the limiting current will depend upon (a) the stirring rate, since if products are made to disperse mechanically the growth of back emf will be delayed, (b) the temperature, since as this is raised the diffusion rate of the dissolved cations will increase, (c) the actual cation, since cations all have different diffusion rates, and (d) the position of the anode since diffusion is, for example, aided by gravity if the dissolution face is downwards and hindered if upwards.

Concentration polarization only becomes important when the dissolution current density i_A approaches i_L, so that the polarization curve shown in Fig. 30 should be extended to give the composite curve shown in Fig. 36.

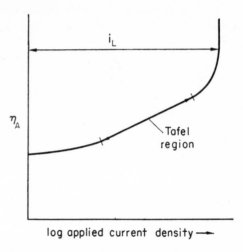

Fig. 36. An anodic polarization curve illustrating concentration polarization.

The smaller the initial exchange current density the larger will be the current density range over which the Tafel equation is likely to apply. Much work on these matters has therefore been done on transition metals.

An electrolyte through which a current is passing will contribute to the overpotential by the factor $i \times R = \eta_R$, where i is the current density and R is the resistance of the path travelled by the current. In good electrolytes η_R is usually small, since the conductivity is high, but in other cases the value, sometimes referred to as the 'IR drop', may be quite high.

Thus, for the comparatively simple case of a metal dissolving in a solution, with no other reactions occurring, the shape of the polarization curve is complicated, being composed of three separate effects: polarization caused by (i) activation energy requirements, (ii) concentration effects, and (ii) resistance effects. A curve exhibiting these three effects is shown in Fig. 37.

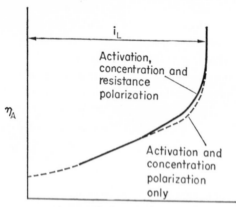

FIG. 37. An anodic polarization curve illustrating concentration and resistance polarization.

2.3. CORROSION KINETICS

Corrosion reactions are more complicated than that indicated above. The anodic dissolution reaction must follow the behaviour indicated above but, in addition, there are other reactions occurring on the same metal surface. The dissolution referred to above was aided by an externally applied potential, with the anodically dissolving specimen connected to an external cathode through a battery. On a corroding metal, anodes and cathodes occur next to each other and any simple corroding specimen includes cathodic sites, on which the reaction rate must at any time be electrochemically equivalent to the anodic reaction rate. Thus for a corroding specimen it can be written:

Anode Total oxidation rate = Total reduction rate *Cathode*

This is a very important observation, since it provides two ways of preventing corrosion. Either the oxidation reaction or the reduction reaction can be hindered. Other things being equal, the effect will be the same. Inhibitors, for example, which hinder such reactions, as discussed in section 3.4, can be either anodic or cathodic or, in some instances, both. The necessary balance between the two reactions is also important in any analysis of a corrosion problem. A corroded metal arises from the existence of a readily reducible species and its removal may reduce considerably the 'aggressiveness' of an environment. Copper alloys, for example, do not normally liberate hydrogen and they exhibit considerable resistance to strong reducing acids provided no oxygen or other reducible species is present.

In many corrosion reactions, in solutions exposed to the air, a common reaction is the reduction of oxygen to hydroxyl ions according to:

$$O_2 + 2\,H_2O + 4\,\text{electrons} = 4\,OH^-$$

where it can be seen that catholytes (regions of electrolyte adjacent to cathodes) tend to be regions of high pH (alkaline). This is also true, of course, if hydrogen evolution is the main cathodic reaction. The normal electrode potential for oxygen reduction is $+0.401$ V and for hydrogen, as we have already seen, is ± 0.000 V.

When a metal starts to corrode in a solution, therefore, there must always be at least one oxidation process, metal dissolution, and one reduction process, oxygen reduction (for example). If the potential of this specimen is measured, the value will be somewhere between the standard potential of the metal electrode and that of the oxygen electrode. It will be a mixed potential. Both reactions will be polarized towards each other and the actual polarization curves will be determined by the considerations outlined above for metal dissolution. The simplest case which will be considered in detail is where both polarization curves exhibit only activation polarization, as shown in Fig. 38. For convenience, the area of the anodic reaction is taken to be equal to the area of the cathodic reaction, although this is a very rare occurrence in actual practice.

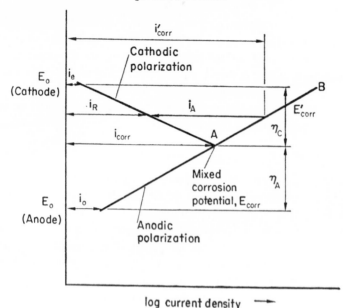

FIG. 38. Combined activation polarization curves for an anodic and a cathodic reaction on a corroding specimen which is anodically polarized.

For the metal dissolution process, equation (17) will apply:

$$\eta_A = b_A \log i_{corr} + a_A.$$

But $\qquad \eta_A + E_0 \text{ (anode)} = E_{corr}.$

Therefore $\qquad E_{corr} = b_A \log \dfrac{i_{corr}}{i_0} + E_0 \text{ (anode)}$

where i_0 = exchange current density for metal dissolution.

For the oxygen reduction reaction at the cathode, a similar equation can be derived:

$$E_0 \text{ (cathode)} - \eta_C = E_{corr}.$$

If i_ϱ = exchange current density for the cathode and b_C = Tafel constant for this reaction, then:

$$E_{corr} = b_C \log \dfrac{i_{corr}}{i_\varrho} + E_0 \text{ (cathode)}.$$

If anodic polarization is applied, the dissolution rate of the metal will be increased, as in the similar ideal case of single metal dissolution, but, in addition, the rate of oxygen reduction, which occurs on the same surface, will be reduced. The actual curve obtained in the corrosion polarization curve will follow the potential according to the line AB in Fig. 38 and with values of i_A of externally applied current density at potential E'_{corr}, etc.

In general $$i_A = i_{corr} - i_R$$

where i_{corr} = the corrosion dissolution current density of the metal and i_R = the reduction current density of the oxygen at potential value E'_{corr}. It is often an oversimplification to describe a corrosion reaction by such an equation, since other reactions may occur simultaneously. These may be due to small traces of impurities in the metal or in the aqueous environment from which they plate out. Small corrosion cells are established. The magnitude of their effect can be considerable. All these possible phenomena, which are often described as local action, will cause a net corrosion loss, which can be designated by a current density term $i_{L.A.}$. This represents additional cathodic reactions causing localized unmeasured anodic dissolution. These reactions can arise from the solution, e.g. metal 'plating out' or some species readily reducible at the potential of the specimen. These reactions may be associated also with impurities in the surface of the metal caused by 'plating out' or from residual alloying elements which do not dissolve but, for example, affect the hydrogen evolution reaction. Local action effects can be minimized by purifying both the environment and the metal under investigation and stringent precautions are taken to achieve this when very accurate work is undertaken.

Hence
$$i_A = i'_{corr} - i_R - i_{L.A.}.$$

However, $$E'_{corr} = b_A \log \frac{i'_{corr}}{i_0} + E_0 \text{ (anode)}.$$

Therefore $$E'_{corr} = b_A \log \frac{i_A + i_R + i_{L.A.}}{i_0} + E_0 \text{ (anode)}.$$

It is clear from this equation that true Tafel curves will not be obtained in corrosion reactions until external polarizing current densities are several orders of magnitude greater than the cathodic corrosion current density and local action current density, since the deviation from the Tafel line occurs to much higher values of external current density than for non-corroding electrodes.

The full polarization curve for a corroding electrode, including concentration and resistance polarization effects, can be written:

$$\eta_A = b_A \log \frac{i_A + i_R + i_{L.A.}}{i_0} - 2.3 \frac{RT}{F} \log \frac{i_L - i_A - i_{L.A.}}{i_L} + i_A R.$$

If a very pure material is investigated, producing no local action, no external current is applied, and de-oxygenated solutions are used so that only hydrogen evolution is possible as a cathodic reaction, then

$$E_{corr} = -b_C \log \frac{i_{corr}}{i_H},$$

where i_H = the exchange current density for hydrogen reduction.

If b_C is measured from the straight line sections obtained for solutions of different pH, and E_{corr} corresponds to the reversible value of the hydrogen electrode at the various pH values employed, then, from E_{corr}, i_{corr} can be calculated. Under carefully controlled experimental conditions Stern[51] obtained very good agreement between calculated and experimental conditions, using pure iron in de-oxygenated H_2SO_4. His results are shown in Table VII.

The electrode potential for an anodic reaction will change with an applied current i_A as predicted by:

$$\eta_A = b_A \log \frac{i_A + i_{corr}}{i_0}.$$

This is only true if (a) no concentration or resistance polarization is present, and (b) no oxygen is present to depolarize the electrode. From this expression has come the term polarization resistance for measuring the corrodibility of a system. The current required to alter

TABLE VII. OBSERVED AND CALCULATED VALUES OF
THE CORROSION OF PURE IRON IN H₂SO₄ IN THE
COMPLETE ABSENCE OF OXYGEN[51]

pH value	0·96	**2·0**
Exchange current density i_H	0·1 $\mu A/cm^2$	**0·11 $\mu A/cm^2$**
Tafel slope, b_C	0·1 volt	0·1 volt
Corrosion potential, E_{corr}	0·203 volt	0·201 volt
Current, calculated from electrical measurements	10·5 μA	11·0 μA
Current equivalent to the corrosion rate obtained by chemical analysis	11·1 μA	11·3 μA

the corrosion potential by 10 mV is measured and

$$\frac{\Delta E}{\Delta I} = \frac{b_A b_C}{2 \cdot 3 i_{corr}(b_A + b_C)}$$

if only activation polarization is operative, where b_A, b_C are the Tafel constants for the anodic and cathodic reactions. Even if b_A and b_C are not known, the corrosion rate can be estimated within a factor of 2 since $(b_A b_C)/(b_A + b_C)$ varies little in magnitude.

For systems where the corrosion rate is controlled by concentration polarization (e.g. oxygen diffusion),

$$\frac{\Delta E}{\Delta I} = \frac{b_A}{2 \cdot 3 i_{corr}} .$$

Thus log $\Delta E/\Delta I$ vs. log i_{corr} gives a straight line of the predicted slope. A reasonable estimate of corrosion rate can therefore be rapidly made without knowing more about the corroding system, other than that it is in an electrochemical phenomenon.

Factors Affecting Polarization Curves

Many factors affect aqueous corrosion reactions, mainly by altering the polarization characteristics of one or more of the electrode reactions. The rate of oxidation in a corrosion reaction must be equal to the rate of reduction, so that the anodic and cathodic currents are equal, as shown in Fig. 39, where the potential/current relationships are formally described as straight lines, although this is rarely found. It has already been pointed out that polarization is a function of *current density*, but with this type of diagram it is possible to illustrate the influence of various factors. Possible effects of altering the ratios of anodic and cathodic areas, which are assumed equal here, are briefly described later.

In Fig. 39 the anode and cathode reactions are polarized to almost the same potential, so that the lines are drawn as intersecting. If a large resistance path exists between anode and cathode sites, then

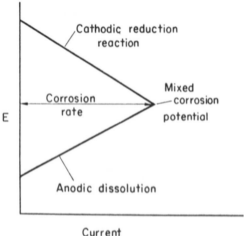

FIG. 39. Potential/current diagram illustrating the polarization of the anodic and cathodic corrosion reactions on a metal specimen immersed in a low resistance medium.

the two reactions will assume potentials that differ by iR, where i is the corrosion rate and R the actual resistance between the sites. This situation is illustrated in Fig. 40. The value of the corrosion potential measured under these circumstances will vary between A and C, depending, *inter alia*, upon the position of the probe electrode.

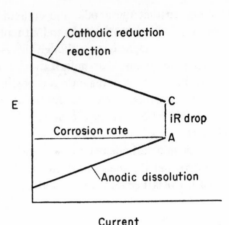

FIG. 40. Potential/current diagram illustrating the polarization of the anodic and cathodic corrosion reactions on a metal specimen immersed in a high resistance medium. There is an iR drop between the sites of the two reactions.

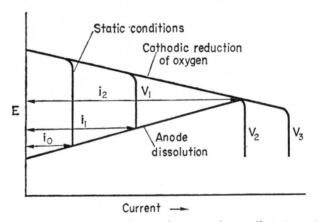

FIG. 41. Potential/current diagram for a metal corroding at a rate determined by the arrival of dissolved oxygen to the metal surface. Stirring the solution at velocity V_1 increases the corrosion rate from i_0 to i_1, while increasing the velocity to V_2 increases the corrosion rate further to i_2. A further increase in corrosion rate to V_3 does not cause any further increase in corrosion rate since oxygen diffusion is no longer the rate controlling factor.

Frequently in aerated solutions, the oxygen reduction reaction is polarized through lack of oxygen, a situation illustrated in Fig. 41. If the solution is stirred or made to flow, the current is increased from i_0, under static conditions, to i_1 at solution velocity V_1, or to i_2 at velocity V_2. If the velocity is increased to V_3 there is no further increase in the corrosion rate, since the diffusion current for the oxygen reduction reaction is no longer the limiting factor.

If the activity of oxygen is increased, e.g. by bubbling oxygen into an unsaturated solution, then the actual starting potential of the reaction is raised, as illustrated in Fig. 42, in accordance with the Nernst equation. The effect is to raise the corrosion rate.

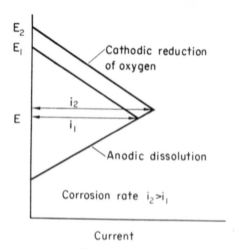

FIG. 42. Potential/current diagram showing the effect of bubbling oxygen into an unsaturated solution. The activity of the oxygen is increased, the potential of the reaction is raised from E_1 to E_2 and the corrosion rate is increased from i_1 to i_2.

If the oxygen reduction reaction is not under diffusion control, stirring or flowing will have little effect upon the corrosion rate. A large supply of oxygen may actually reduce the corrosion rate, if it helps to restore an oxide film to the metal, an effect that has been reported for both iron and zinc in neutral solutions (cf. section 3.4).

The actual solubility of oxygen in an electrolyte is an important factor in determining corrosion rates. The solubility of oxygen, with change in concentration of a sodium chloride solution, is shown in Fig. 43 and the corrosion of iron in aerated salt solution shows a similar variation. It is unfortunate that sea water, corresponding approximately to a 3% solution of sodium chloride, represents the maximum on this graph.

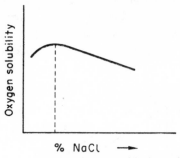

FIG. 43. The change in solubility of oxygen with change in concentration
of sodium chloride solution.

Hydrogen peroxide, which readily dissociates into water and oxygen, will keep the oxygen content of an electrolyte high and prevent concentration polarization. Nitric acid, which is readily reducible, provides an easy alternative reaction which polarizes only to a very small extent, resulting in a high corrosion rate. The effect of both materials, which act as cathodic depolarizers in different ways, is shown in Fig. 44.

Anodic depolarizers are less common, since they must interfere with the metal dissolution reaction. Sulphur lowers the anodic activation polarization of iron and therefore flattens the polarization curve. If copper is added to steel, the precipitation of cuprous sulphide helps to remove this harmful effect. Contact with external sulphides is undesirable for the same reason. Steel structures under the earth often suffer severe corrosion through the action of bacteria, *Desulphovibrio desulphuricans*, which reduce sulphate ions to sulphide ions under anaerobic conditions (cf. section 3.2).

The potential difference between the cathode and anode reactions will be enlarged if the activity of the metal ions is lowered since, in accordance with the Nernst equation, this will lower the anode potential. This effect is readily achieved by complexing agents. Copper,

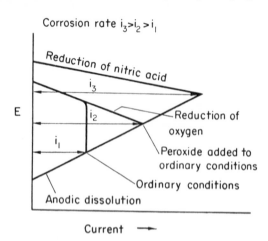

FIG. 44. Cathodic depolarizers. For a metal whose corrosion rate i_1 is dependent upon the diffusion of oxygen to its surface, the addition of hydrogen peroxide to the solution will increase the corrosion rate to i_2 because the peroxide readily supplies oxygen to the solution and prevents diffusion control. The addition of nitric acid provides an *alternative* cathodic reduction reaction and increases the corrosion rate to i_3. The magnitude of the effects of these additions is very dependent upon the specific conditions and upon the concentrations employed.

for example, will not corrode in de-aerated sulphuric acid since the only possible cathodic reaction, hydrogen evolution, occurs at potentials below that of copper dissolution. If HCN is added to the acid the copper forms a complex anion:

$$Cu^+ + 4\,HCN \rightleftharpoons Cu(CN)_4^{3-} + 4\,H^+.$$

The reaction is almost complete and the equilibrium constant K, $[Cu(CN)_4^{3-}]/[[Cu^+][CN^-]^4]$, is very large. $[Cu^+] < 10^{-21}$ g ions/litre.

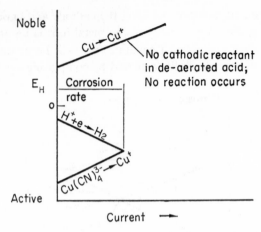

Fig. 45. The effect of cyanide ions in causing the corrosion of copper in de-aerated sulphuric acid by forming complexes.

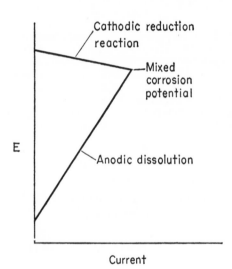

Fig. 46. The reaction under anodic control. Changes in the anodic polarization curve will alter the corrosion rate more than the same proportionate changes in the cathodic polarization.

E for Cu^+/Cu = $+0.52$ V. From the Nernst equation:

$$E = 0.52 + \frac{RT}{F} \ln a_{Cu^+},$$

and since $a_{Cu^+} < 10^{-21}$, $E \approx -0.72$ V (or more negative).

The effect is illustrated graphically in Fig. 45. Since hydrogen evolution is a possible cathodic reaction, the copper will corrode.

If a corrosion reaction can be represented by Fig. 39, then it is said to be under mixed control, since changes in either the anodic or cathodic reactions may have quite a large effect upon the corrosion rate. If the situation is like that in Fig. 41, the reaction is said to be under cathodic control since alterations in the anode line will scarcely affect the intercept. In Fig. 46, the reaction is under anodic control.

2.4. HYDROGEN EVOLUTION CORROSION REACTIONS

In all the discussions above it was assumed that oxygen reduction was the predominant cathodic reaction. This is often true, but there is a second cathodic reaction that is commonly found, which is the evolution of hydrogen. This reaction is often the main one (a) in acid solutions, (b) in solutions containing complexing agents, and (c) in the corrosion of very active metals.

Whether it is possible for a particular metal can be determined by examining the Pourbaix diagram for the metal. The metal must go into solution at a potential that is active with respect to hydrogen evolution. Any metal electrode will meet the requirement if the activity of the metal is sufficiently low. This underlines the significance of complex formation and low solubility as discussed in section 2.3.

The hydrogen evolution reaction has been the subject of considerable investigation. There are different ways of considering it, since there is a series of steps in the reaction and it is not always clear which one is the important rate determining step. Initially there is proton discharge. The hydrogen ion which is hydrated is converted to an adsorbed hydrogen atom:

$$H_3O^+ + \text{electron} = H_{ads} + H_2O. \tag{20}$$

This atom can form a hydrogen molecule by combining with a second atom formed nearby. For this to occur, surface migration of the atoms towards each other must take place. This implies that the atoms are not strongly adsorbed. Molecule formation is hindered by the dissolution of some of the hydrogen atoms into the metal which removes the atoms from the surface. Thus molecules are formed in competition with atom dissolution:

$$H_{ads} + H_{ads} = H_{2_{ads}}. \tag{21}$$

This gas molecule may form a bubble—perhaps with other molecules—and escape.

Alternatively, secondary discharge may occur on the same site as the first, in which case molecule formation occurs without the need for surface diffusion:

$$H_3O^+ + H_{ads} + \text{electron} = H_{2_{ads}} + H_2O.$$

The important rate determining step in this reaction for a large number of metals is the reduction step in equation (20). The rate at which it occurs is related to the heat of adsorption of hydrogen for each metal. For strong adsorption, the exchange current density, i_0, is large and for weak adsorption it is small. Some values are shown in Table VIII. For some metals with large heats of adsorption, i.e. Pt and Pd, the reaction in equation (21) becomes rate determining at high rates of discharge.

The significance of variations in i_0 for each metal lies in the effect of polarization upon hydrogen discharge. The change in potential required to discharge hydrogen at a certain rate will depend upon i_0 as is shown in Fig. 47. Some values of hydrogen overpotential are shown in Table IX.

The ease with which hydrogen is discharged from a metal surface depends very much upon the metal in question. This observation

TABLE VIII. SOME APPROXIMATE
EXCHANGE CURRENT DENSITIES
FOR THE HYDROGENE ELECTRODE
AT 25 °C

	i_0 (A/cm^2)
Pt	10^{-2}
Pd, Rh	10^{-4}
W, Co, Tn	10^{-5}
Fe, Au, Mo	10^{-6}
Ni, Ag, Cu, Cd	10^{-7}
Sn, Al, Be	10^{-10}
Zn	10^{-11}
Pb, Hg	10^{-13}

Source: J. M. West, *Electrodeposition and Corrosion Processes*, Van Nostrand, London (1970).

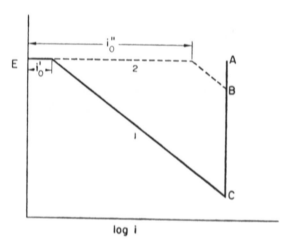

FIG. 47. Schematic polarization curves for hydrogen evolution on two metal surfaces. Metal 1 has a small exchange current density i_0' and requires a polarization of AC to discharge hydrogen at the same rate as on metal 2, which has a larger exchange current density i_0'', which requires a polarization of AB. Metal 1 has a high hydrogen overpotential and metal 2 has a low hydrogen overpotential.

TABLE IX. HYDROGEN OVERVOLTAGE
REQUIRED TO DISCHARGE HYDROGEN
AT DIFFERENT RATES ON SOME
METALS IN N H_2SO_4

Metal	Current density (A/cm²)		
	10^{-4}	10^{-3}	10^{-2}
Pt bright	–	0·02	0·07
Au	0·12	0·24	0·39
Fe	0·22	0·40	0·56
Ni	0·20	0·56	0·74
Ag	0·30	0·48	0·76
Cu	0·35	0·48	0·58
Pb	0·45	0·52	1·09
Zn	0·55	0·72	0·75
Hg	0·60	0·78	0·93

Source: U. R. Evans, *The Corrosion and Oxidation of Metals: Scientific Principles and Practical Applications*, E. Arnold, London (1960).

must be borne in mind always when considering corrosion reactions where the principal cathodic reaction is hydrogen evolution. In particular, consideration must be given to the composition of the metal surface which can be expected to change. Such events will depend upon the composition both of the environment and of the alloy. From the environment, cations may 'plate out'. Their effect upon the corrosion rate will depend upon the amount deposited and upon the hydrogen overpotential of the deposited metal in relation to the metal in question. From the metal, impurity cations that do not go into solution or do not stay in solution because they are noble with respect to the metal will accumulate upon the dissolving surface and will affect the corrosion rate in a manner that will be dependent upon their hydrogen overpotential in relation to the metal in question.

Examples of deposition from the environment and accumulation for the metal can now be cited.

In Fig. 48 the corrosion rates of zinc and iron in acid solutions are shown graphically. If platinum ions are in the solution or are added directly, they 'plate out' on both metals. The hydrogen overpotential for platinum is much less than that of either zinc or iron. The intersection of the anodic dissolution curve and the hydrogen evolution reaction on the deposited platinum surface occurs at a much larger

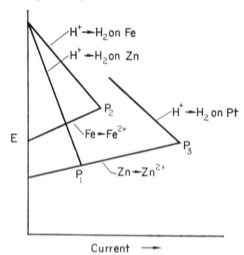

Fig. 48. The effect of the addition of a platinum salt upon the corrosion of zinc and iron in acid solutions[52]. The platinum is deposited on the metal surfaces and provides areas of low hydrogen overpotential. The corrosion rate of zinc increases from the value at intercept P_1 to P_3 while the corrosion rate of iron increases from the value at intercept P_2 to the value at the intercept of the anodic curve with the cathode curve (not drawn).

current density than when hydrogen evolution is occurring only on either zinc or iron. The corrosion rate is therefore increased. If the element deposited had a higher hydrogen overpotential than the two metals, then their corrosion rate would be diminished. Segregation within an alloy, or the presence of a second phase, may also affect the corrosion rate in acid solutions. Iron containing graphite or cementite corrodes at higher rates than pure iron, since both of these phases have lower hydrogen overpotentials than iron itself.

In Fig. 49 the changes in the corrosion rate of some zinc alloys and zinc in N/2 H_2SO_4 are drawn. The corrosion rates of the alloys change, depending upon the hydrogen overpotential of the solute element, provided that it 'plates out'. Mercury has a higher hydrogen overpotential than zinc and the corrosion rate of the dilute zinc–mercury alloy therefore diminishes with time, as the mercury accumulates on the surface. The corrosion rates of the other alloys increase as the specific solute elements accumulate on the surfaces.

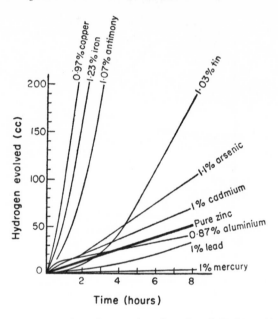

FIG. 49. The corrosion of some zinc alloys in N/2 H_2SO_4 with time. The changes result from the accumulation of impurities on the alloy surface which have different hydrogen overpotential characteristics[53].

In Fig. 50 the corrosion rates of the same alloys in aerated N/10 KCl solution are plotted. All corrode at approximately the same rate[54]. In neutral solutions, the principal cathodic reaction is oxygen reduction and this reaction is not dependent upon the surface composition of the zinc alloys.

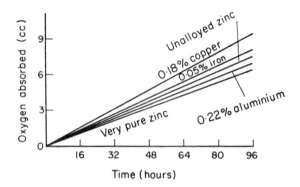

FIG. 50. The corrosion of some zinc alloys in aerated N/10 KCl vs. time. The predominant cathodic reaction is the reduction of oxygen and this is not affected by any accumulation of impurities[54].

The environmental and alloy impurity effects are quite general and can be expected to occur until equilibrium is established between the metal dissolution and impurity deposition reaction, such that the activities of the various species present correspond to a common value of electrode potential that can be calculated from the Nernst equation. Consequently, quite small impurity levels may exert effects which may become apparent only after quite long periods of time. Even the pure zinc in Fig. 49 will contain some impurities and provided that these accumulate, they will have an effect upon the overpotential.

Hydrogen evolution is a very complicated reaction made up of several steps. Each of these steps may have several stages, so that the reaction, which has been quite widely studied, is really like a long-chain reaction. Furthermore, the rate determining stage is not always the same single one. Impurities may either increase or decrease the evolution rate, since the different surface that they provide will affect one or more of the reaction stages. Whether this effect concerns the initial adsorption of hydrogen ions on the metal surface, the formation of hydrogen atoms, the proton mobility or molecule detachment is not always clear and it is unlikely that impurities affect only one of these steps. Arsenic and sulphur, for example, appear to delay the atom–atom combination step of equation (21) and thereby increase the probability of hydrogen solution into the metal. On ferrous alloys, in particular, hydrogen embrittlement and blistering are frequently caused by environments containing compounds of these species.

Values of hydrogen overpotential are difficult to establish. On some metals, e.g. aluminium, most of the surface is likely to be covered with an oxide film. On others, surface roughness will vary. Many measurements have been made using a mercury drop, on which the surface must be atomically flat. Consequently, values of overpotential (being the overpotential required to produce a stated rate of hydrogen discharge) must be accepted with care and circumspection. The overpotential varies from metal to metal and so does the exchange current density of the reaction, which is related to the thermionic work function of the metal electrode. The larger the work function, the greater is the exchange current density[55], and this indicates the importance of the electron transfer step in the discharge mechanisms.

The experimental problems of determining hydrogen overpotential values and Tafel relationships are considerable, since minute traces of impurities can affect the reaction. Even an alteration in the surface tension of the electrolyte, by a trace of detergent, may be important since it will encourage the formation of gas bubbles.

2.5. THE CORROSION OF ALLOYS

The effects of alloying additions upon the corrosion of alloys can be divided into three groups:

(a) Changes in anodic dissolution kinetics of the solute metal.
(b) The effect of impurities upon local action cells, which has already been discussed.
(c) Changes in grain boundary attack caused by segregation and heat treatment.

Changes in anodic dissolution kinetics must be considered for both single and multi-phase alloys. In the former, the more active component will tend to be dissolved at a faster rate than the less active. For example, brass is readily dezincified in many solutions as rapidly as

it is exposed to the solution. The dissolution of copper thereby becomes rate controlling, although it must be pointed out that there
are several steps in this reaction. For 85/15 brass, for example, the
oxidation of cuprous ions to cupric ions at the alloy surface is the
slowest step[56]. The general result is a surface enrichment in the less
active element and a gradual reduction in the dissolution rate, provided that the atom ratio of noble/active elements is above a value
which is probably specific to each system, that the alloy is not covered
with a film, and that the alloy is homogeneous.

It has been argued[57] that for an alloy X, Y

$$i_{\text{alloy}} = i_Y \left(1 + \frac{C_X}{S}\right)$$

where i_{alloy} = dissolution current density of the alloy, i_Y = dissolution current density of the solvent, C_X = concentration of the solute,
and S = specific gravity of the alloy, the necessary assumption being that
the dissolution rate of X is very much larger than that of Y at the same
potential. This equation is derived from adding the current densities
of the single metals on the assumption that each contributes a proportion to the alloy current density. This proportion is the product
of the current density of the pure metal and the fraction of the alloy
area that the metal occupies. The equation appears to hold for limited
ranges of composition of some alloys.

In heterogeneous alloys much depends upon the relative areas of
the exposed phases. It has been found both experimentally and
theoretically[58] that at any potential E, for a galvanic couple of components A and B,

$$i_T = i_A f_A + i_B f_B$$

where i_T = total current for the two components A and B, f_A, f_B =
area fractions for the components and i_A, i_B are the separate dissolution rates for the alloys at the same potential. If the polarization
curve for each phase of a binary alloy is known, the polarization curve
for the alloy can be calculated from this equation, since the area
fractions can be calculated from the volume fractions. If an active

phase is finely dispersed in a noble matrix, it will be dissolved out and further dissolution of the alloy will be determined by the noble matrix. If the active phase is continuous, or present in a sufficiently large proportion, this will not happen. Surface enrichment of the more noble component should occur therefore only in dilute heterogeneous alloys, but as the dispersion becomes finer this possibility is increased. In this respect, a homogeneous alloy can be regarded as a heterogeneous alloy reduced to the finest possible dispersion and surface enrichment becomes the rate determining factor.

If all solid state interactions are neglected, then at a potential E the current density for each component in a homogeneous alloy is the same as that for the pure metal. Assuming dissolution to be uniform, the ratio A/B is the same in the solution as in the alloy. Therefore

$$\frac{N_A}{N_B} = \frac{z_B i_A f_A}{z_A i_B f_B}$$

where N_A/N_B = mole fractions of A and B, and z_A, z_B are the oxidation numbers (valencies) of the ions[59].

Therefore
$$f_A = \frac{z_A N_A i_B}{z_A N_A i_B + z_B N_B i_A}.$$

Therefore
$$i_T = \frac{i_A i_B (z_A N_A + z_B N_B)}{z_A N_A i_B + z_B N_B i_A}. \tag{22}$$

The components are not of course independent of each other thermodynamically and there must be some interaction. If this can be ignored and if the dissolution process is under activation control, then its rate is determined not by the corrosion potential E but by the anodic overpotential η. The condition of constant overpotential can be achieved by moving the polarization curves[59] of the separate components along the potential axis until their corrosion potentials coincide with that of the alloy, after which the polarization curve for the alloy can be calculated from equation (22). The equation will account for surface enrichment, but it will fail if the alloy contains an excess of active component. If surface enrichment requires the removal of large numbers of active atoms, it becomes difficult to establish. The small number of noble atoms will probably be detached along with the large number of active atoms. The limiting case for the dissolution of a homogeneous alloy is the normal case for the dissolution of a heterogeneous alloy and vice versa.

It must be emphasized that the calculation of anodic polarization curves and the dissolution rates of alloys is not a simple matter. Much remains to be done on this subject. If any element in an alloy segregates to the grain boundary, then the difference in concentration may result in a potential drop between the grain boundaries and grain interiors which is sufficient to develop corrosion currents of some magnitude, or it may produce locally active regions. The effect can be very marked, even though the alloying element may be present in very small amounts, e.g. as an impurity in a nominally pure metal. Grain boundary effects of this type are widespread. Two examples will be cited.

The intergranular corrosion of aluminium in alkali is caused by the presence of iron as a residual impurity in the grain boundaries. This probably acts as a very good local cathode for hydrogen evolution. The same effect is found in acid solutions.

A much-quoted example of heat treatment resulting in grain boundary attack is the deterioration of 18 Cr–8 Ni stainless steels, after heating to certain temperatures such as arise during welding, etc. Chromium carbide is precipitated in the grain boundaries and areas immediately adjacent to them are thereby depleted of chromium and become active with respect to the grain interiors. If aqueous conditions exist, a cell is set up as illustrated in Fig. 51, with the denuded areas anodic to the rest of the surface. The large cathode area provides anodic control and attack is very severe, leading to intergranular failure or pitting. Associated with welding treatment, this source of trouble is referred to as 'weld decay' and the steel is referred to as being 'sensitized'. If stronger carbide formers are also alloyed with the steel, e.g. niobium or titanium, then these precipitate as carbides, leaving the chromium in solid solution and eliminating this form of attack. Such a steel is termed 'stabilized'. The problem is reduced

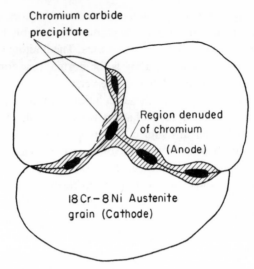

Chromium carbide
precipitate

Region denuded
of chromium

(Anode)

18 Cr – 8 Ni Austenite
grain (Cathode)

Fig. 51. Weld decay. Intergranular corrosion caused by the precipitation of chromium carbide. This denudes the adjacent regions of protective chromium, which therefore become anodic to the bulk of the material.

also by using low carbon grades of austenitic stainless steels. There is evidence[60] to suggest that a second cause of weld decay arises from the segregation of phosphorus and sulphur to the austenitic grain boundaries. Austenitic alloys based upon nickel can also be sensitized and care must be exercised in welding procedures in order to lower the probability of intergranular corrosion.

Grain boundary attack is not confined to metals containing impurities and alloys that have some form of segregation. When a pure metal is etched, attack is greater at the grain boundary than within the grain itself. A groove forms between the grains which scatters perpendicular light from the microscope and thereby provides a ready indication of the boundary that a metallographic study requires.

Grain boundaries are active regions, and it is not surprising that the disarrayed layers of atoms are more readily attacked than the regularly arrayed atoms of the grains. It is also to be expected that any region of disarray will be a site for preferential attack. Emergent dislocations, subgrain boundaries, twin boundaries, etc., are all subject to increased corrosion attack. Much depends, however, upon the film covering the metal. A metal whose oxide is removed by the solution will have its whole surface attacked. With care and the correct choice of etchant, dislocation sites can be revealed. These often take the form of etch pits and much experimental use has been made of them. The number of pits and number of emergent dislocations do not necessarily correspond, but this is not usually important. A metal whose oxide is not removed by the solution will suffer attack where the oxide breaks down, and these sites will not always correspond to the underlying metal surface. Pits on aluminium will form on sites determined partly by weak points on the film and partly by environmental and geometric factors. If the metal is plastically strained, however, pits form along the line of the emerging slip plane where the film has been fractured[61].

When a bare metal crystal is attacked by a solution, the dissolution rate varies according to the crystallographic orientation, quite apart from the influence of crystal imperfections. Some reagent mixtures produce pits with flat faces that correspond to a specific crystallographic plane. Once this is known, surface orientation is easily ascertained. On aluminium, for example, a mixture of fuming nitric acid, hydrochloric acid, hydrofluoric acid and butyl cellosolve[62] produces pits with [100] sides. The crystallographic nature of the pitting of aluminium in chloride solutions has also been shown[63]. Attack on very thin foils was shown to produce narrow 'tunnels' which changed direction orthogonally. The dissolution rate appeared to be constant, since the application of anodic polarization only broadened the penetration front and did not increase the penetration rate.

2.6. GALVANIC CORROSION

Galvanic corrosion may arise when two dissimilar metals are in contact in an aqueous environment. The potential difference between them will initiate attack at a corrosion rate that is largely dependent upon the surface reactions of the two metals. Galvanic potentials are made use of, practically, in batteries. In the Daniell cell, containing solutions of unit activity, the standard potential of copper in contact with cupric ions, $+0.34$ V, and zinc with zinc ions, -0.77 V, produces a potential difference of 1.1 V. In most examples of galvanic corrosion the situation is less simple, since the aqueous environment will not usually contain ions of unit activity and the metal surface is largely covered with an oxide film. Under these circumstances reference to the electrochemical series must be made only with care. For example, aluminium occurs below zinc and should be anodic to it. This is not found experimentally. Zinc protects aluminium covered with an oxide film and is anodic to it. Even analogy with cells is dangerous, since the cathodic reaction in galvanic couples is usually either oxygen reduction or hydrogen evolution, not metal deposition. Copper does not normally liberate hydrogen and in a galvanic cell the cathodic reaction is the reduction of dissolved oxygen. In many situations where the environment is isolated from the air, e.g. a central heating installation, the initial oxygen in solution is rapidly consumed and not replaced, so that the galvanic current falls to a negligible figure. Under such conditions copper and steel can readily be used together. In aerated cnvironments, however, such a couple would cause attack on the steel.

Surface films affect the efficiency of an electrode. A bare metal is a much better cathode than one covered with an oxide, which, apart from its possible interference with hydrogen evolution, puts an additional resistance in the electrochemical circuit. The stability of the oxide in the solution is therefore of some importance. The potential difference developed between aluminium and stainless steel is about the same as that developed between aluminium and copper. In the first case the cathodic stainless steel is covered with a highly protective oxide film which has a low conductivity. The galvanic current between the two metals is therefore comparatively small. In the second case, however, the oxide on the cathodic copper is readily reduced (as described in section 1.8 on electrometric reduction) and oxygen reduction occurs at a high rate at the efficient bare metal surface. In these two cases where the reaction is under cathodic control, the efficiency of the cathode determines the rate of corrosion, and this is a common situation.

Since the diffusion of oxygen is frequently the rate determining factor in aqueous corrosion, large ratios of cathode area to anode area will frequently result in intense galvanic attack. Such effects are most likely to occur at joints of structures joined together by a different metal. They can be avoided by plating, e.g. with cadmium or zinc, provided that the second couple is a safer one. Since iron oxide, like the oxides of copper, is readily reducible, steel is often cadmium plated if it is to be used with aluminium. Metals can be insulated from each other, one coated with the other, painted, etc., to avoid dangerous couples. Much depends upon the specific service conditions.

Galvanic corrosion can occur without the two metals being initially in actual contact. Water containing one metal may cause corrosion of a second if the first 'plates out' on the second. If soft acid water (containing carbonic acid) flows through copper pipes and into a galvanized tank, any copper taken up from the pipes is likely to be deposited:

$$Cu^{2+} + Zn = Cu + Zn^{2+}$$

and the efficient copper cathodes will produce intense local attack on the zinc coating. Similarly, if two metals are insulated from each other, corrosion may still occur if the corrosion products of the cathodic metal are deposited on the anodic metal and reduced to the bare metal, e.g. Fe and Al.

Some pairs of metals can change polarity. Zinc is anodic to iron and is widely used as a coating, a process referred to as 'galvanizing'. Under aqueous conditions zinc corrodes preferentially protecting the iron. Loose flocculent $Zn(OH)_2$ maintains anodic conditions at the zinc surface but above about 60 °C a hard compact ZnO layer is formed instead, and this is cathodic to the iron. If protection is required some other anodic material must be used, e.g. magnesium. Magnesium is a very widely used sacrificial anode material in oil tanks, ships, etc. (cf. section 3.2).

Crevice corrosion can arise under galvanic corrosion conditions, but the term also includes all forms of corrosion which are peculiar to crevice-like conditions. These may be created by dust particles on a surface which may be hygroscopic, joins between two riveted plates, etc. Much crevice corrosion arises from the effect of the exclusion of oxygen. Some metals which are highly resistant in the presence of oxygen, e.g. titanium and stainless steel, can suffer badly from this type of attack. Intelligent design can help reduce this type of attack by eliminating sites where moisture may collect.

2.7. ATMOSPHERIC CORROSION

Metals are always covered with a very thin oxide film. Usually this quickly reaches a limiting thickness, but there are many cases, particularly in polluted atmospheres, when metals and alloys continue film growth at ambient temperatures. The degree of humidity and the contents of the atmosphere are the two important factors that determine the extent of atmospheric corrosion.

There is usually a critical humidity corresponding to condensation conditions beyond which the amount of atmospheric corrosion increases considerably. The more polluted the atmosphere, the more aggressive the condensed film will be.

When exposed above pure water, a metal surface is covered with a layer of moisture that is very thin, being one molecule thick at a relative humidity of 60% and two at 90%[64]. Should any hygroscopic material be present in this layer, then it will become very much thicker and the conditions are then aqueous. Small particles of such substances including carbon and ash dust therefore become nucleating sites for corrosion attack. Polluted atmospheres frequently contain sulphur and its oxides. The gases are dissolved in the moisture and some metal surfaces catalyse the oxidation of sulphur from four- to six-valent. The end product consists largely of sulphuric acid. Many tarnish films formed in the atmosphere contain large proportions of sulphate ions. The sulphur and sulphide products that are not oxidized may react with the metal surface to form sulphides which are either partially oxidized to sulphate or else stimulate the anodic reaction, as explained above in section 2.2, and increase the corrosion rate. Sulphides are more defective than oxides and they therefore grow at faster rates.

The concentration of chlorides in the atmosphere diminishes rapidly from the high value at a sea coast, unless there is a prevailing wind on shore. Chlorides are very prone to initiate localized corrosion attack on what are usually resistant surfaces, e.g. aluminium and chromium plate, resulting in the creation of pits which can become quite deep, leading to penetration.

Since there are pollution and humidity variables and since some surface reactions are conditioned largely by the air-formed oxide and the reaction of the metal with the anion, it is difficult to generalize about reactions of metals and alloys in atmospheres. A few examples will instead be quoted.

Iron

Rusting is nucleated under humid conditions by particles and the rust spots spread outwards irregularly until they meet each other, after which time the rate of rusting remains fairly constant. Sometimes rust spreads, not as an expanding patch but along a very narrow path. These paths are comparatively straight, but they change direction when they approach each other and never cross. This very thin line form of rusting is termed filiform corrosion and it is illustrated in Fig. 52. The original saltspot absorbs water and after rust is precipitated the site is covered with a membrane of hydrated ferric oxide. This will burst at some point, perhaps from volume increases within the membrane, and fresh attack will occur. As alkali accumulates around the sides over a longer period of time than at the 'head', attack proceeds in an approximately straight direction. This changes when the path comes near the cathodic liquid of another path.

The mechanism of atmospheric rusting in industrial atmospheres has received much attention. In particular the role of SO_2 which is of considerable importance has been much examined. It appears to take

Lines of rust

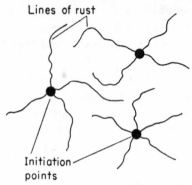

Initiation points

FIG. 52. Filiform corrosion.

part in a cyclical series of reactions, since it has been shown[65] that each SO_2 molecule can result in the formation of 15–40 molecules of rust, the amount depending upon the season of exposure. Two proposed cycles have been put forward. Firstly, the *acid regeneration cycle* which involves a sequence of reactions following the oxidation of SO_2 to SO_3 and its dissolution in water:

$$Fe + H_2SO_4 \rightarrow FeSO_4 \rightarrow Fe_2(SO_4)_3 \rightarrow FeO \cdot OH + H_2SO_4.$$

Secondly, the *electrochemical cycle* in which iron is gradually incorporated into rust ($FeO \cdot OH$):

$$Fe = Fe^{2+} + 2e$$
$$Fe^{2+} + 8\, FeO \cdot OH + 2e = 3\, Fe_3O_4 + H_2 \cdot O$$
$$3Fe_3O_4 + 0 \cdot 7\, SO_2 + 4 \cdot 5\, H_2O = 9\, FeO \cdot OH.$$

In this series of reactions eight molecules of rust become nine molecules.

Recent work by Evans[66] supports the electrochemical cycle. Acid regeneration is not ruled out, but is considered to occur very slowly by comparison with the electrochemical cycle. In the acid regeneration cycle, insoluble basic sulphate is thought to occur. This will result in the gradual removal of Fe^{2+} and SO_4^{2-} from the liquid. This removal provides a possible explanation of why the proposed recycling equations do not give rise to an infinite amount of rust.

Aluminium

Aluminium is attacked when first exposed, but the rate soon becomes negligible. Under conditions where oxygen is excluded, e.g. crevices and dust particles, local sites of intensive corrosion may be initiated.

Nickel

Nickel was formerly used as a protective covering for steel, but it becomes covered with a light-coloured film. When exposed to sulphurous atmospheres it develops a haze that can be wiped off at first but requires abrasion after longer periods of exposure. At first the haze consists of free sulphuric acid and nickel sulphate, but it is later transformed to basic nickel sulphate.

Nowadays the nickel layer is plated with a thin layer of chromium which provides a bright finish because of the protective nature of the oxide film on chromium. The effectiveness of the composite layer in protecting iron is largely dependent upon the thickness of the nickel layer.

Some typical atmospheric corrosion rates are shown in Table X[67].

TABLE X. RATES OF CORROSION OF ORDINARY IRON AND STEEL IN THE OPEN AIR AT DIFFERENT LOCALITIES PRIOR TO 1940[67]

Location	Rate of rusting	
	Ingot iron (mils per year)	Mild steel (mils per year)
Khartoum	0·06	0·02
Singapore	0·59	0·46
Llanwyrtyd Wells (Wales)	2·22	0·92
Congella (South Africa)	2·97	1·44
Motherwell (Scotland)	3·58	2·02
Woolwich (London)	3·9	1·84
Sheffield	5·09	4·25

1 mil = 0·001 in.

2.8. PASSIVITY

In section 2.2 the dissolution process of a metal was described as an oxidation process of the general form, $Me \rightarrow Me^{z+} + z$ electrons. From the Pourbaix diagrams it is clear that the electrolytic oxidation of a metal can take other forms too. If the metal can be oxidized to an oxide that is stable in the electrolyte, then the metal is rendered passive (or is passivated). It is a condition that usually requires strong oxidizing conditions. Iron, for example, is heavily attacked by dilute nitric acid, yet in concentrated nitric acid it is inert, since a very thin protective passive film is formed. In that condition, iron behaves like a much more noble metal than it actually is, e.g. it does not displace copper from a copper sulphate solution.

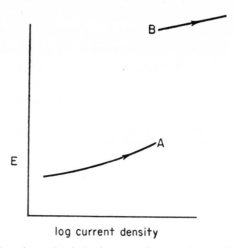

FIG. 53. Experimental polarization curve for a specimen exhibiting passivity, determined galvanostatically.

If an anodic polarization curve is drawn for a metal that exhibits passivity, the shape is similar to that drawn in Fig. 53. When the externally applied current density is increased beyond a value known as the critical current density, the potential "jumps" and oxygen is evolved from the metal surface. The metal at potentials above A is

passivated and is covered with an oxide film. The range of potential
A to *B* cannot be easily investigated under galvanostatic conditions.
To make measurements within that region, it is necessary to control
the potential rather than the current. Since this is an important difference, the device for controlling potential, known as a potentiostat,
will now be described.

One method to ensure that the potential across a cell is kept constant is to connect a resistance very much lower than that of the cell
across its electrodes in parallel. Under such conditions, changes in
the cell resistance will not affect the potential across the compound
resistance. This arrangement is the basis of the classical potentiostat[68]. A typical example is shown in Fig. 54.

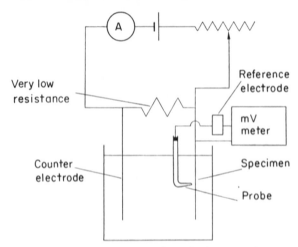

FIG. 54. A typical arrangement of the classical potentiostat.

The total resistance, *r*, of the cell is given by:

$$r = r_a + r_c + r_e$$

where r_a, r_c and r_e are the resistances of the anode process, cathode
process and of the electrolyte respectively. The electrode resistances,
r_a and r_c, will vary with the current through the cell in a manner
dependent upon the polarization characteristics of the reactions.

In potentiostatic investigations it is necessary to measure current through r_a as a function of the potential difference across it. If the total series resistance of the rest of the circuit outside the cell is R_c, then the total resistance of the whole circuit R, excluding r_a, is:

$$R = R_c + r_c + r_e.$$

Since the current through R will affect the potential difference across r_a, R must be kept as low as possible in order to minimize interference with the measurement of r_a.

For a specimen exhibiting passivity the anodic polarization curve is like that shown in Fig. 55, *ABCDEF*. The line *BC* represents unstable conditions that cannot be maintained and if the potential is raised

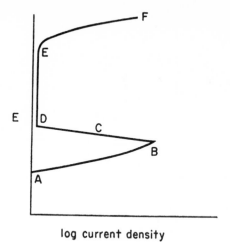

log current density

FIG. 55. Anodic polarization curve for a specimen exhibiting passivity, determined potentiostatically.

into that region it will automatically rise to position *D*. The potential regions *BD* cannot therefore be investigated. The range of this region depends upon the value of R, compared with the value of r_a. In the extreme case, under galvanostatic conditions, R is very much larger than r_a and the trace of the polarization curve follows *ABF*, which is the same as that in Fig. 53.

Electronic instruments are now available that perform the same function as the circuit shown in Fig. 54. The potential difference between a specimen and a probe reference electrode is controlled so that any deviation in potential from a chosen value is corrected via a dc amplifier whose output feeds a third auxiliary electrode (often made of platinum). This third electrode can be either anodic or cathodic to the specimen, depending upon the required correction. The potential setting control can be motor driven and the potential and current output automatically recorded. In this way, curves derived under the same conditions can be readily obtained and usefully compared.

The anodic polarization curve, obtained under potentiostatic conditions, for a metal that exhibits passivity differs considerably from Fig. 30. It shows that a large reduction in current density occurs once the critical current density is exceeded. The potential at this point is often referred to as the primary passive potential, E_{pp}. The fall in current density may be several orders of magnitude and the establishment of the stable lower value corresponds to the existence of a very thin passive film which has a comparatively high electrical resistance. The function of the current is to maintain the film which is slowly dissolving. If the film dissolves slowly, then this corresponds to a highly resistant alloy. The corrosion rate is dependent, therefore, upon the solubility of the film. The nature of the film, its defect structure and the nature of the anions that may be incorporated into it, all help to determine film solubility. This can be quite high for some metals and certain situations so that the rate of metal loss will then be large. Perhaps such situations should not be described by the term 'passive', which tends to be associated with the synonyms inert or unreactive. It should be noted that there is no commonly accepted definition of passivity. Some workers would allow the term to include any anodic polarization behaviour where there is a reduction in current density, however small, upon making the potential noble. Perhaps some other term is required for the general phenomenon of film formation and the term 'passivity' should be reserved for when the film that is formed dissolves at a rate below a certain value. In this text, passivity is implied generally to be associated with a slowly dissolving film. In other texts care should be exercised to see if the same implication is intended.

Passive films have some electronic conductivity. They are not insulators. Consequently, only comparatively small potentials can be maintained across passive films (about 1 V). If higher anodic potentials are applied, then the potential on the solution side of the film will rise and anodic reactions corresponding to the noble potential values will occur. The most common reaction is the evolution of oxygen. If the film formed is an insulator, then much larger potential differences can be maintained across it and the film will grow. Under these conditions the metal is anodized, a process considered further in section 3.6. For a passive metal then, the oxide is maintained at a constant thickness. If the oxide is dissolved too rapidly, the potential across the thinning film will increase the rate of passage of cations from the metal lattice into the oxide, while, should the film become too thick, this rate will be slowed down since the potential 'pull' across the thickening film will be weaker[69].

Passivity cannot be easily established nor maintained in the presence of aggressive anions, e.g. Cl^-. As the concentration of these ions is increased, the critical current density is increased, the primary passive potential is raised, the current density under the condition of passivity is enlarged and the passive potential range is lowered, as illustrated in Fig. 56. The explanation for this behaviour probably lies in the high charge density on the chloride ion and its easy migration. In the passive potential region, it competes with the oxidizing species and becomes incorporated into the film. This produces lattice defects, thereby reducing the resistivity of the oxide.

If a metal is anodically passivated and the controlling potentiostatic circuit is then removed, the potential of the specimen falls, as illustrated in Fig. 57. The metal then becomes active again. The potential corresponding to the re-establishment of active conditions is called the Flade potential. It is pH-dependent:

$$E_F = E_0 - 0.059 \text{ pH}$$

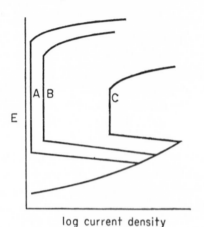

Fig. 56. The effect of chloride ion upon the anodic polarization curve for a metal that exhibits passivity. *A*, No chloride ions present. *B*, A low concentration of chloride ions. *C*, A high concentration of chloride ions.

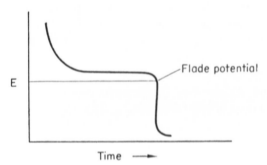

Fig. 57. The change in potential of a passivated specimen when the anodic polarization is removed.

where E_F = Flade potential at the pH under examination and E_0 is the is the Flade potential at pH = 0.

The Flade potential is associated with the dissolution of the protective passive oxide film.

A similar relationship is found for the establishment of passivity:

$$E_{pp} = E_{pp}^a - 0.059 \text{ pH}.$$

E_{pp} and E_F are usually very close to each other and frequently their values correspond to those thermodynamically calculated for the specific oxides that are thought to provide passivity. For iron, however, where the passive layer consists of γ-Fe_2O_3 and, possibly, other oxides, these potentials have values not corresponding to known oxide potentials. The reason for this is unknown, but it may be associated with the defect structure of the passive oxide[70]. The observation that for the reactivation of iron the $E/\log i$ plot has a slope of 59 mV has been similarly interpreted[69]. This association of defect structure and dissolution rate of oxides is an important one. For example, γ-Fe_2O_3 is stable in acids yet in contact with iron in acids it readily dissolves by an electrochemical mechanism:

Anode: $Fe \rightarrow Fe^{2+}_{aqueous} + 2e$

Cathode: $2\,Fe^{3+} + 2e \rightarrow 2\,Fe^{2+} + \square\,(anion\ vacancy)$

$\rightarrow 2\,Fe^{2+}_{aqueous}$

This is called reductive dissolution, since it produces cations from a cathodic reaction. It is only possible in oxides containing cations of variable valency. It represents a further possible cathodic reaction that may enhance corrosion processes.

It must be emphasized that passivity can also occur in regions outside those indicated by the Pourbaix diagram, where the film is only thermodynamically metastable.

Iron, in common with other transition metals, exhibits passivity in the presence of acids with oxyanions, e.g. $SO_4^=$, NO_3^-, $CrO_4^=$, $TeO_4^=$, etc. The actual mechanism of establishment of passivity is not generally agreed. There has to be some adsorption, perhaps as shown in Fig. 58, followed by a desorption and oxide growth. There are a number of steps required and which of them is absolutely critical is not clear. The electronic orbital arrangement within the atoms may determine the first critical adsorption step[71]. The stable oxide lattice structure may be important in controlling the defect concentration, within certain limits, so that the film has the necessary electronic conductivity.

The cause of passivity has long been a subject of considerable discussion. It has been suggested, for example, that an adsorbed layer

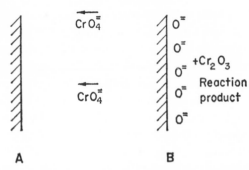

FIG. 58. The passivation of an iron surface (A) by chromate ions. The surface is oxidized (B) and the film that is formed incorporates the chromic oxide that is produced through the reduction of the chromate ion.

of oxygen atoms—chemisorbed—renders all the active dissolution sites inoperative. Subsequent film growth would then be a consequence and not a cause of passivity. The nature of the adsorption process has not been fully elucidated, but vacancies in the d-band orbitals appear to be of significance and it can be emphasized that transition metals in particular exhibit passivity. The existence of a highly charged layer of oxygen atoms may cause the migration of cations into the layer so that it may consist of two separate species, a situation similar to that described in section 1.3 for the initial stages of oxidation.

Recent work[72] has shown that both adsorption and film growth are important and that both occur. An Fe–24 Cr alloy passivated only a few millivolts above E_{pp} exhibited a completely reversible polarization curve when the potential was lowered into the active region, indicating monolayer formation and removal. If it was polarized more than a few millivolts above E_{pp}, then there was a hysteresis effect in making the alloy active again, corresponding to the reduction of a multi-layer film. The longer that the alloy was thus polarized, the longer was the delay in removing the film, a result that suggested film growth under the conditions of polarization.

Values of critical current density are only really comparable if determined on surfaces of the same roughness, yet it is usually the geometric area that is divided into the current. When iron is anodically polarized in sulphuric acid, the anolyte becomes saturated with both ferrous and sulphate ions, which precipitate as $FeSO_4$ crystals. At a later stage the passive oxide is formed, the true current density then being as high as 10–20 A/cm^2.

It is important to emphasize here that passivation implies the establishment of thin oxide (*ca.* 10 nm (100 Å)) films of low solubility. These reach a limiting thickness which is not the same under all conditions. Passivity exists over a definite potential range and all electrodes exhibiting passivity have anodic polarization curves like Fig. 55, exhibiting a sharp potential dependent fall in current density at the onset of passivity. Many metals react with their environments to form insoluble films by direct chemical reaction, which is not potential dependent, and these must not be confused with those exhibiting passivity. Lead, for example, is resistant to sulphuric acid since the sparingly soluble sulphate forms on the lead surface and cuts off further reaction. This reaction is not potential dependent and the anodic polarization curve would not therefore show a sharp reduction in current density. Lead cannot therefore be described as passive in this environment.

In order for an alloy to be used in the passive condition the *mixed corrosion potential*, E_{corr}, must occur in the passive range. This can be achieved if the situation, drawn in Fig. 59, is achieved. Under such circumstances, the alloy is described as 'self-passivating', since no externally applied current is necessary and the essential condition for this is that at E_{pp} $i_{cathodic} > i_{critical}$. If alloying lowers i_{crit} (perhaps by altering the adsorption process), then an active alloy may become passive if the above condition is achieved. For a passivatable alloy, additional cathodic reactant may cause an active metal to become passive. An example of each can be cited.

Iron–chromium alloys show an increasing tendency to passivate as the chromium content is increased. The critical current density required for the establishment of passivity in de-aerated neutral solutions falls as the chromium content is raised to 12%, beyond which it is constant at $2\mu A/cm^2$[73]. This is a very low value. If a cathodic reaction occurs on an Fe–12 Cr alloy at a higher rate, and if its potential is relatively

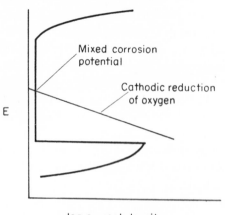

log current density

FIG. 59. The self-passivation behaviour of stainless steels in aqueous environments containing oxygen. For a Fe–12 Cr alloy in a de-aerated neutral solution, when the cathodic reaction is the evolution of hydrogen, the critical current density is 2 μA/cm[73].

noble, then the alloy will be passivated, since the critical current density will be exceeded and the mixed corrosion potential will be in the passive range. This is also true for stainless steels in aerated solutions and for the same reason: $i_{cath} > i_{crit}$. Thus ordinary stainless steels are self-passivating, where moisture and oxygen are present, and exhibit relatively noble potentials, as would be expected from the anodic and cathodic polarization curve intersection shown in Fig. 59. Self-passivation is a very important feature of a corrosion resistant alloy, since whenever the protective film is broken it re-forms immediately.

An example of self-passivating alloys being developed through increasing i_{cath} at E_{pp} is the Ti–Pd alloy system. These alloys show excellent resistance to strong acids in which titanium metal would corrode. Ti–Pd in boiling 10 wt% HCl corrodes at 0·002 in/year approximately, while Ti metal corrodes at 0·2 in/year[74]. In practice, the material is supplied as titanium with a very thin porous layer of palladium. The palladium accumulates on the surface, for reasons described in section 2.4. The hydrogen overpotential of palladium is much lower than that of titanium and i_{cath} increases until the potential E_{corr} goes into the noble range. From examination of Fig. 59, it should be clear that acid resistance is only possible in this way on those metals or alloys where E_{pp} is *active* with respect to the potential for hydrogen evolution, which will depend upon the pH of the solution, as explained in section 2.2. The mechanism for Ti–Pd appears to be one of dissolution, followed by palladium deposition, since frequent changes of the environment delay the establishment of the fully protective surface. This alloy constitutes a useful inert anode and is employed in cathodic protection installations.

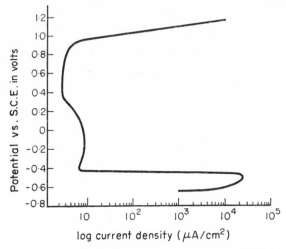

FIG. 60. Anodic polarization curve for chromium[75].

A similar series of acid resistant alloys can be made, based upon chromium, but these do not resist oxidizing acids. The explanation for this difference lies in the anodic polarization curve for chromium, which is shown in Fig. 60[75]. The increase in current density at 900 mV is called transpassivity and represents the dissolution of chromium in the hexavalent state as chromate ions.

When chromium is alloyed with small amounts of platinum, palladium and other noble metals, its corrosion resistance to non-oxidizing acids is markedly improved. The dissolution of chromium leads to a gradual enrichment of the alloy surface by the noble element. The hydrogen overpotential on noble elements is very low and the exchange current density is comparatively large. The cathodic polarization curve of the alloy is gradually flattened until the critical current is exceeded and the alloy goes passive. In oxidizing acids a second cathodic reaction is possible, viz. the reduction of the anion, which occurs at potentials above that of the hydrogen evolution reaction. For chromium in nitric acid, the cathodic reaction for the reduction of NO_3^- occurs in the potential region near to the onset of transpassivity. The additions of noble elements raise the potential further, since the exchange current density of the reaction is larger on the noble metal surface. The mixed corrosion potential of the alloy reaches the transpassive region, where chromium dissolves as Cr^{6+} and

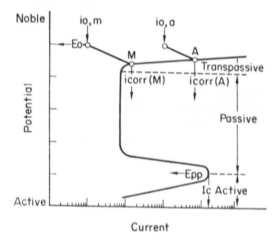

FIG. 61. Effect of noble metal additions to the corrosion reaction of chromium. In non-oxidizing acids the metal is passive. In oxidizing acids it is raised into the transpassive region[75]. $i_{0,m}$ and $i_{0,a}$ are the exchange current densities for the cathodic reduction reaction on the metal and alloy surface respectively. $i_{corr\,(M)}$ and $i_{corr\,(A)}$ are the corrosion rates of the metal and alloy respectively.

the dissolution rate of the alloy is increased. This is shown diagrammatically in Fig. 61[75]. If the noble element is dissolved by the nitric acid, then this increased attack will not occur. Additions of noble elements to stainless steels have been employed[76] to increase their resistance to non-oxidizing acids by maintaining the mixed corrosion potential in the passive region.

2.9. METHODS OF TESTING

Some methods used in studying aqueous corrosion phenomena have already been described in sections 2.2 and 2.3. In this section it is intended to describe further methods and to indicate the interpretation of the experimental observations.

Experimental arrangements for studying aqueous corrosion range from the very simple to the very complex. All are valuable and the choice must be governed by the objective of the proposed investigation.

A large amount of valuable information can often be obtained from a careful visual examination. An outstanding example is the salt drop experiment described in section 2.1. The distribution of corrosion sites can yield considerable data to the trained observer. This subject is a large one and it is difficult to generalize, but in acidic and strongly alkaline solutions attack is frequently widespread, since the oxide film may be dissolved, a general situation observable from many Pourbaix diagrams. In fairly neutral solutions, considerable variation in corrosion distribution is found. Since the oxide is stable, attack will be initiated by weak spots, determined both by the oxide and by the underlying metal. These will include, *inter alia*, areas of damaged oxide, or irregular covering, arising perhaps from the contours of the metal, e.g. abrasion lines and grain boundaries, and structural factors in the metal, e.g. emergent screw dislocations and slip steps. The further spread of attack will depend upon many factors. If the oxide is very stable, attack may concentrate at the initial site. Oxygen availability is an important variable and so is the physical nature of the corrosion product, which can vary considerably. It may be flocculent, freely porous to the solution (including oxygen), compact ($PbSO_4$), or intermediate, like a membrane which prohibits access of fresh solution to the corroding site. As the trapped solution becomes depleted of oxygen, the pH falls, resulting in an increased rate of attack. On aluminium, which has a very stable film, the small anodic sites are surrounded by very large cathodic areas and this is a dangerous situation. The result is the production of deep pits. In hard waters, the precipitation of insoluble hydroxides and carbonates, at the adjacent cathodic regions, tends to keep the pits small and thereby promote deep penetration. In soft waters, the local alkaline regions tend to spread the pits over a wider area and cause less penetration. On iron, in contrast, the film is less stable, since it dissolves at a higher pH on the acidic side than Al_2O_3. Furthermore, reductive dissolution occurs readily. Attack is therefore widespread all over the metal.

This is a very simple presentation and in practice there is much variation. Often, areas adjacent to corrosion sites are protected, e.g. filiform corrosion or peripheral areas of pits which are acting as cathodes and are to some extent protected. The membranous nature of rust is dependent upon the phases present. These will be ferrous or ferric oxides in various states of hydration and their occlusive effect will vary considerably. This whole subject is of tremendous importance. It is never sufficient merely to know what forms, since its physical state may often have overriding control of a reaction.

The quantitative measurement of aqueous corrosion can often be determined from weight loss measurements. An example is shown in Fig. 62 for steel, immersed in water and covered with sand[77]. The sand probably interferes with the oxygen reduction reaction. It also forms a hard crust with the rust. If such measurements are to be meaningful, attack must be uniform. Weight loss measurements on the pitting of aluminium will not indicate the dangerous nature of the penetrating attack. Instead, in pitting investigations, statistical measurements of numbers of pits or depth of pits are often used. In this type of investigation considerable care is required in order to make a safe assessment. If the number of pits of increasing depth gradually falls off, then experimentally measured depths are not likely to indicate the maximum depth, unless a very large number of

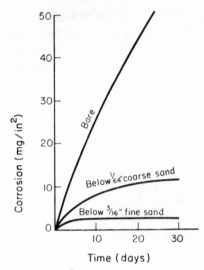

Fig. 62. Corrosion of steel plates under water. Effect of sand layers[77].

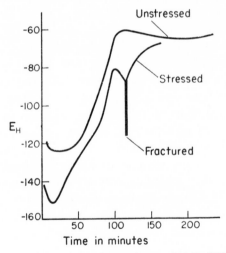

Fig. 63. Potential/time curves for specimens of 18 Cr–8 Ni steel immersed in 42 wt% $MgCl_2$ solution at 154 °C.

specimens are examined. If, however, the number of pits of increasing depth falls off rapidly, then experimental measurements will probably include the maximum depth. Differentiating between these possibilities is not easy. Data on deepest pit depth will be useful if the second case applies, where there is some limiting velocity of attack, but may be far less meaningful if the first case applies[78].

Electrochemical investigations of corrosion have already been mentioned in section 2.2. Changes in the corrosion potential under open circuit conditions can be informative. In Fig. 63, for example, the changes in corrosion potential of 18 Cr–8 Ni austenitic stainless steel wires in 42 wt% MgCl$_2$ solutions, boiling at 154 °C, are shown for both stressed and unstressed specimens. At first, the solution attacks the metal through pores and weak points in the film. The potential shifts downwards in the active direction. Within these gaps the solubility products of the various hydroxides are soon exceeded and they precipitate out, stifling the corrosion attack. This sequence is indicated by an upward movement of the potential. The precipitation gradually leads to a lowering of pH within the pores, which produces soluble corrosion products and film breakdown. On unstressed specimens, this leads to pitting, which is not accompanied by further changes in the potential. On stressed specimens, stress corrosion cracks (see section 4.3) form, which propagate through the alloy, resulting in failure. This sequence is accompanied by a fall in potential, with a sudden 'kick' upon failure, which is accounted for by the temporary increase in exposed bare metal.

Polarization has been mentioned throughout this chapter. Many factors contribute to the shape of a polarization curve. The direct electrochemical components were described in section 2.3. The experimental arrangement of the electrodes, particularly the probe, have also been considered. The effect of impurities in causing significantly large local action currents has been briefly described in section 2.3.

Many other factors require consideration when starting experimental work. A few examples can be mentioned. Polarization in acidic solutions requires prior oxygen removal, since otherwise the total cathode current will be divided between two reactions and the potential may be shifted to such noble values that hydrogen cannot be evolved. Other species exhibiting similarly inconvenient redox potential values must be excluded too, for the same reason and also because should they 'plate out' they then alter the hydrogen discharge characteristics. Sometimes solutions are electrolysed before use in polarization experiments in order to remove impurities. Oxygen removal can be done in several ways, the most common being to saturate the solution with de-oxygenated nitrogen or hydrogen prior to use.

Since the standard electrode potential of iron is -440 mV, hydrogen is readily evolved from iron during polarization experiments in acid solutions. The most active part of the anodic polarization curve is obtained by polarizing at constant current density and measuring the true anodic current density from the weight loss. This is demonstrated in Fig. 64. The corrosion potential A is lowered to D by cathodic polarization, with an externally applied current density CD. At this

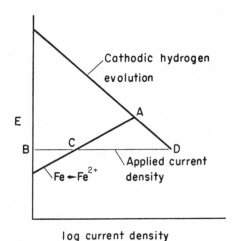

FIG. 64. Determination of the true anodic polarization curve for iron in de-aerated acid solutions. The anodic dissolution current density BC is found by measuring the weight loss of a specimen maintained at a constant potential by the applied cathodic current density CD. Each point requires a separate specimen.

potential, iron is dissolving at the rate indicated by *BC* and hydrogen is being evolved at the rate indicated by *BD*. *BC* is measurable by weight loss, which is then converted to the equivalent current density. By using different values of applied current densities for a set of specimens, a true anodic polarization curve can be obtained.

It has been emphasized in section 2.2 that polarization is a function of current density. Different experimental curves will therefore be obtained from surfaces of the same geometric area, if the roughnesses of the two are considerably different. In a lot of work, controlled pre-etching, following a standardized abrading treatment, is used with much success in order to obtain reproducible roughnesses.

The time taken to produce a polarization curve affects the final result. In anodic polarization, dissolution may alter the surface roughness. Under galvanostatic conditions the current density will be altered and therefore the potential will shift. Under potentiostatic conditions the current required to maintain the potential will alter. Apart from anodic dissolution roughening the area there may be accumulation of anodic products, or other surface reactions, which may interfere with the determination of the polarization curve. A change in conditions at any one reading will affect later settings of the circuit. Therefore the speed of measurement is important. The direction in which the potential is moved can also affect the results. Many different experimental arrangements are employed, including the use of electronic devices to change the potential over the whole range under investigation in milliseconds. These are complex matters which are the subject of much discussion and investigation. Comparison between polarization curves obtained under different conditions can frequently be very difficult and care must be exercised in attempting such comparisons.

Since the value of the potential determines the current density that emanates from an electrode, it might be expected that the potential would be plotted on the axis of the abscissa with the current density as the ordinate. That this has not been commonly done in the past can be explained probably by the widespread employment of galvano-static experimental arrangements, as described above, in which the current is set and the potential then measured. With the increasing use of potentiostatic arrangements, there is an increasing tendency to plot polarization curves in what would appear to be the more correct method. Since the large majority of workers still use the older plotting arrangement, that practice has been used in this text.

Many other experimental arrangements have been used, to achieve various ends. Concentration polarization can be important, even at low current densities. To minimize it, polarization experiments are often done with rotating cylindrical electrodes. This is probably more satisfactory than stirring of the solution mechanically, since conditions are easily reproduced. For some work single crystals are necessary. Most metal single crystals appear to develop faces of simple orientation in equilibrium with solutions, independent of the orientation of the face initially exposed to the solution. These will usually be the most slowly dissolving faces and are often close-packed faces. Electrode assemblies have been the subject of much study, since the correct positioning of electrodes is not always a simple matter. Minute glass electrodes are sometimes employed to measure pH changes occurring during corrosion reactions.

Many other techniques have been used to study phenomena in aqueous corrosion experiments. The choice is largely determined by the aim of the investigation. New problems arise continually and there is a constant need for the ingenious imaginative approach based upon a sound understanding of the principles involved.

CHAPTER 3

The Protection of Metals

3.1. INTRODUCTION

In Chapter 1 several methods for improving the oxidation resistance of metals were described, while in Chapter 2 resistance to aqueous corrosion was noted under some circumstances.

Under oxidation conditions there is not much freedom of action. One is forced to use the correct material, or at least one that has a useful length of life. In some cases a coating can be used with an otherwise unsuitable alloy. The dangers here if protection is lost are obvious. Since a number of high melting point metals (commonly referred to as refractory metals) have poor oxidation resistance, high temperature coatings[79] have been developed employing oxides, silicides, aluminides and other compounds.

Under aqueous corrosion conditions it is rarely possible to use non-corroding noble metal coatings, for economic reasons. The problem, therefore, is to combat the intrinsic corrodibility of such metals and alloys as are in general use. There are several methods for protection against aqueous corrosion which are derived from electro-chemical principles. Others perform the obvious task of separating the metal from the environment. The success of the latter depend upon the chemical or electrochemical resistance of the protective layer and its mechanical properties.

The choice of method is not easy. Quite apart from the simple fact that the methods of protection are by no means universally applicable, choice is partly governed by the actual environmental conditions and partly by economic considerations. The latter include not only the initial cost of application but also the replacement of corroded parts and, in some cases, renewal of the protecting medium. These are frequently the determining factors and the practising corrosion engineer will use economic parameters as familiarly as scientific data in determining his choice. No attempt is made here to cover the costing methods of corrosion protection, but they must never be forgotten.

The electrochemical methods of protecting a metal can best be understood by considering the appropriate Pourbaix diagram—for example, that of iron, which has been illustrated in both the complete and the extremely simplified versions in Figs. 33 and 34. Iron will not corrode when its potential and its environmental pH come into the region of immunity, where the metal is more stable than any other possible state. To achieve this condition, the metal must be polarized so that its potential is lowered from some value, representing freely corroding conditions, to a value somewhere below its normal electrode potential. This is the basis for cathodic protection, which is discussed in section 3.2. An alternative method of electrochemical protection is to ensure that the iron electrode is within the region of *passivation*. This will require polarization to make the potential more noble if the pH value of the surrounding medium is approximately between 2 and 9. Between 9 and 12 approximately, iron is either immune or passive, while at pH values < 2 a passivation region is not found. This type of protection, often called anodic protection, is described in section 3.3. Unlike cathodic protection, it is not applicable over the whole range of pH and it can be 'lost' if the film is destroyed, e.g. by the presence of chloride ions. Cathodic protection can be maintained in any environment provided that there are no extraneous effects, e.g. direct chemical attack on the metal. The distinction may be recalled here that immunity is a region where corrosion (i.e. electrochemical attack) *cannot* occur, while passivation is a region where corrosion *does not* occur.

3.2. CATHODIC PROTECTION

Corrosion under aqueous conditions is an electrochemical phenomenon and has been discussed at length in section 2.2. The dissolution of the metal occurs as an anodic process. If the potential of a corroding object is lowered to the reversible potential of the anodic reaction, then the anodic dissolution will stop, since the dissolution rate is equalled by the metal deposition rate (exchange current density) at this value of the potential, so that there is no net loss. Essentially the whole surface of the object is then providing sites only for the cathodic corrosion reaction, which will usually be either hydrogen evolution, oxygen reduction or both together. This is the basis of cathodic protection.

The situation is illustrated in Fig. 65. Under freely corroding conditions the specimen has a mixed corrosion potential E and is corroding at a rate equivalent to i_{corr}. If cathodic polarization is now employed so that the potential of the specimen is lowered to E_1 by the externally applied current i_1, then the specimen will be partially protected since the corrosion rate will have been reduced from i_{corr} to

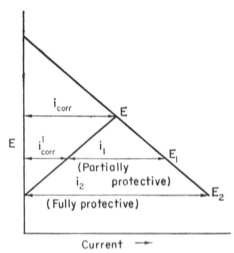

Current →

Fig. 65. Cathodic protection. Applied cathodic current i_1 reduces the corrosion rate from i_{corr} to i'_{corr}, partially protecting the metal, while applied cathodic current i_2 produces complete protection.

i'_{corr}. If the externally applied current is increased to i_2 so that the potential of the specimen falls to E_2, the reversible potential of the anodic reaction, anodic dissolution will be stopped. The specimen is then under full cathodic protection.

This is a simple presentation of cathodic protection, but in practice it is more complicated. There is the time factor to consider (as there is in all polarization experiments). Suppose a metal Me of valency z is cathodically protected in an environment that contains no Me^{z+} ions. If the mixed potential of the metal is lowered to the reversible potential of the anode reaction, $Me \rightarrow Me^{z+} + z$ electrons, as represented by E_2 in Fig. 65, then under equilibrium conditions, as defined in section 2.2, the metal will be in contact with the environment containing Me^{z+} ions at unit activity. Since the bulk of the environment is free of Me^{z+} ions, a concentration gradient will be established and Me^{z+} ions will diffuse away from the metal surface. The metal must then dissolve in order to maintain the correct concentration of Me^{z+} ions in contact with its surface. Under potentiostatic conditions the protective current density will increase. Under galvanostatic conditions, which are used more often in practice, the corrosion rate will increase. The corrosion rate will depend upon environmental factors that govern the diffusion of the cations. In practice the potential to which the specimen is lowered is usually below the reversible electrode potential of the anode reaction. It is taken to some value, corresponding, in accordance with the Nernst equation, to a lower activity of Me^{z+} ions in contact with the metal.

The environment is important, since it may react with the products of the cathodic reaction. Iron will often become covered with hydroxides and carbonates of magnesium or calcium, since these ions are often present in water, which also usually contains CO_2, and the final product is a rusty, chalky mixed deposit. This will provide considerable protection to the metal. Under these conditions the polarizing current required for full protection, i_2, will fall, its main function then being to repair breaks in the protective scale. Even if interaction of this kind does not occur, the increase in alkalinity produced by the cathodic reaction will tend to provide protection even if the potential is not lowered to the value E_2 in Fig. 65. Reference to the appropriate Pourbaix diagram will show this. As the surface of the metal becomes covered with a protective film, the value of the required polarizing current will fall. Excessive alkalinity, however, is dangerous since many metals form soluble oxyanions under high pH conditions.

In practice there are two ways of providing the necessary polarization for protection of a metal: (a) by employing naturally occurring galvanic corrosion, using a less noble metal which is gradually corroded, or (b) by using an impressed current, with either inert or expendable anodes.

Cathodic protection is usually concerned with the protection of ferrous materials, since these constitute the bulk of objects that are buried in the earth or immersed in water, e.g. pipelines, foundations, piers, jetties, ships, etc. Magnesium is widely used throughout the world as a sacrificial anode. It is usually in the alloyed form with 6% aluminium, 3% zinc and 0·2% manganese, since these help to prevent film formation, which would reduce the dissolution rate of the metal. The protective current efficiency is always less than 100%, since cathodic hydrogen is evolved from the alloy, i.e. the magnesium corrodes. Aluminium alloyed with 5% zinc is also used, but the potential difference for this alloy against iron is much smaller than for magnesium alloy. It is near to that of zinc metal itself, which is also used for protection when suitably alloyed to prevent anode film formation, which is associated with iron, a common impurity in zinc. The choice between anode material is a difficult one. In soils or other environments of low conductivity, a large potential difference is necessary, since the iR drop between electrodes will be substantial, whereas in environments of high conductivity, a low potential drop may be more economical to use. The positioning of the electrodes, the throwing power of the environment, i.e. its ability to provide adequate current density to all parts of a surface, and the polarization characteristics of the electrodes are all important variables. If electrodes are to be buried, it is quite common practice to surround them with neutral, porous conductive material, referred to as backfill, if the ground is unacceptable for some reason, e.g. aggressive to the anode.

The use of impressed current methods for cathodic protection permits much greater control over the system and is often cheaper to apply than the use of sacrificial anodes, which, of course, need regular replacement.

Inert anodes that are used include platinum and stainless steel, which both evolve anodic oxygen. Carbon anodes, including graphite and agglomerates of carbonaceous material, are also used but they tend to be consumed through the chemical formation of CO_2. Platinum is used in sea water in the form of a very thin electrodeposited layer on titanium. The latter would suffer corrosion under impressed anodic currents, but the presence of platinum raises the potential into the passive range (cf. section 2.8). The result is a substantial strong anode with a large area of platinum.

If inert electrodes are buried, the backfill must be porous to allow for the escape of oxygen.

Consumable electrodes used with impressed current are often large pieces of old rusty scrap iron or steel. They are eventually consumed and must then be replaced. If this requires extensive digging, then the total cost of the installation is considerably increased. This, of course, is also true for consumable electrodes used without impressed current.

Power for impressed sources comes from ac in industrial countries after transformation and rectification, while for the protection of pipelines in the less developed countries, long running diesel motor generators are used. Other sources, e.g. windmills, are used where local conditions make them advantageous.

Cathodic protection is applied extensively to buried ferrous structures, particularly pipelines, and to ships where anodes are often built into the hull.

Pipeline corrosion is a complex subject. Apart from the aggressiveness of soils, which varies considerably, there are two other unusual features which complicate cathodic protection, (a) long line corrosion, and (b) bacterial corrosion.

A pipeline may run through soils whose oxygen permeability differs considerably. That portion that passes through heavy anaerobic clays can become anodic with respect to that portion that passes through sandy light porous soils which are full of oxygen. This is an example of a differential aeration cell. Where oxygen is scarce and permeability poor, the oxygen reduction reaction will be polarized to a greater extent than where oxygen is readily available. The difference in mixed potential of the metal in the two soils represents a cell within itself. The anode and cathode parts may be miles apart and this type of attack is called long line corrosion. In applying cathodic protection over the whole length of a pipeline that is likely to suffer this type of attack, it is clear that the amount of protection required will vary according to the soil.

Bacterial attack arises in anaerobic soil. These often contain bacteria, called *Desulphovibrio desulphuricans*, which reduce $SO_4^=$ ions to $S^=$ ions. Under the anaerobic conditions the corrosion potential is depressed since there is no oxygen available to depolarize the cathode. Hydrogen evolution is the cathodic reaction and this proceeds at a slow rate. The sulphide ions depolarize this reaction very considerably and they also appear to decrease the activation polarization for the dissolution of ferrous ions. The result is a savage attack on the pipe, with the formation of a non-protective scale of ferrous sulphide.

The bacteria flourish in neutral soils. Cathodic protection is effective against them in many circumstances, since it creates alkaline regions in contact with the metal and in some cases it helps form a protective scale. Low pH also reduces the activity of the bacteria, but acid soils are often aggressive, since hydrogen liberation takes place more easily. Bactericides are generally effective and can be added either as solutions or be impregnated into sand. Alkaline treatment of soil is sometimes undertaken to combat these bacteria.

In practice, cathodic protection is rarely used alone. The current required for full protection would usually be far too great and apart from necessitating a more expensive electrical installation it would often produce deleterious side effects, e.g. excessive alkalinity. Instead, cathodic protection is used in conjunction with some form of covering. The protective current required is therefore small and serves to protect any exposed parts of the metal surface.

The protective coating used depends upon the conditions of

exposure. There is a considerable choice and it is not possible to compare one with another very readily, since opinions differ, experience varies and fresh developments occur regularly.

Painting is described in section 3.7. It is used with cathodic protection, but care has to be taken, since some paints do not withstand alkaline conditions, e.g. linseed oil-based paints are prone to suffer from 'alkaline peeling'.

In laying pipes the protective procedure depends upon the aggressiveness of the soils. Pipes are separated from the worst soils by lining the trench with brick and rubble. In most cases pipes are wrapped with cloth saturated with bitumen, asphalt or coal-tar, after being primed. The wrapping is often fibreglass or one of several flexible plastics. Enamelling can be used instead of hard-setting coatings, including cement. Choice is governed by such factors as: Is the ground subject to varying loads, e.g. under a road. Is there a varying water table. Does the soil contain hard rocks that rest against the pipe, etc.? It is worthwhile to note that careful handling is important when laying pipes, in order to avoid damage to protective coats.

Cathodic protection in soils can exacerbate corrosion in adjacent unprotected buried metal parts. It must be taken into account when designing systems of protection. Attack can also occur from stray currents, from whatever source they come, e.g. electric trains. Short-circuiting is usefully employed here, but is not always the answer. While dc will nearly always cause trouble, only some metals suffer from stray ac. Aluminium is an example and some aluminium-containing brasses. This reaction probably arises from the rectifying property of aluminium oxide.

3.3. ANODIC PROTECTION

In cathodic protection the rate of the anodic dissolution reaction is reduced by lowering the value of the mixed corrosion potential of the specimen to the reversible potential of the anodic reaction, preferably with a generous allowance against possible error. Even under these conditions there will be a small corrosion rate, but this is not objectionable provided that it is uniform. In section 2.8 it was shown that when a metal undergoes an active/passive transition there is a considerable reduction in current density to a very low value. This is made use of industrially and the protection of a metal by maintaining it in the passive condition is called anodic protection.

The difficulties in using it are considerable. Whereas cathodic protection can be used to protect many metals immersed in any medium that conducts, e.g. solid or liquid, anodic protection is used at the moment only for protecting complete sections of chemical plant, which must be made from metal that can be passivated in the contents of the plant. This immediately limits its use. Furthermore, it is potentially dangerous, since if protection breaks down at any point and is not immediately re-established, dissolution at that point will be very rapid since the gap in the film provides a path of low resistance while the metal is anodically polarized.

The use of anodic protection requires careful planning of the chemical plant. The plant must be monitored so that any loss of protection is rapidly brought to the attention of the operator. There must then be provision for the re-establishment of protection. This may only require a local boost of anodic current, but in the worst case it may entail the immediate draining of the whole plant.

Anodic protection does not stand up to aggressive anions. Chloride ions destroy the passive film and their concentration must therefore be kept low except in the case of titanium, which can be passivated in hydrochloric acid. The throwing power of an electrolyte under anodic protection is very good[69] so that only a comparatively small number of electrodes is required to maintain protection once it is established. Under conditions prior to passivation, the throwing power is poorer, a fact to be borne in mind in designing protection installations.

Industrially, anodic protection has been used to reduce the corrosion of mild steel in contact with sulphuric acid, and with aqueous fertilizer solutions containing ammonia and ammonium nitrate temperatures as high as 95 °C have been successfully employed[80].

Anodic protection requires very little power, can be applied to common structural metals that exhibit passivity, e.g. plain carbon and stainless steels in many common environments, is readily controllable and measurable and requires no expensive surface treatment of the metal. It uses the natural phenomena of reactions between a container and contents. It is an elegant method and its use seems likely to spread as the problems of monitoring are overcome.

3.4. INHIBITION

Cathodic and anodic protection both attempt to employ electrochemical measures to reduce the corrosion rate of metals by external polarization. Another general approach is to separate the metal from the corroding medium by coating the surface of the metal. The several methods of achieving this separation are described in sections 3.5–3.7. There is one other approach, which is to reduce the 'aggressiveness' of the environment towards the metal by small additions that hinder corrosion reactions, either by reducing the probability of its occurrence or by reducing the rate of attack or by doing both. The effect of the addition agents in reducing corrosion is called inhibition. A distinction can be made between the two main types of solution which may require inhibitor treatment. With one type the pH is within the neutral/alkaline range, while with the second type the pH is within the acidic range and the two types correspond respectively to a situation where a film may be formed on the metal, a process which the inhibitor promotes, and to a situation where the metal surface is bare and the inhibitor itself must provide a protective adsorbed layer. Neutral inhibition is considered first.

Inhibition covers a very wide field of anti-corrosion measures. Although a lot of work on the subject is concerned with preventing the attack of specific chemicals upon certain metals during manufacture, transport or storage, the bulk of inhibition in practice is directed towards water treatment. The problem in this latter case is at least threefold:

(a) To reduce the contamination of water by metals.
(b) To lower the corrosion rate and thereby increase the life of tubes and other containers, e.g. motor car radiators.

(c) To reduce the failure of boilers and ancillary equipment in steam generators.

In descaling operations that employ acid it is also necessary to reduce attack on the metal.

Before discussing the methods that are both generally and specifically employed, it is necessary to look at the mechanism of aqueous corrosion, in particular the reaction of iron with water, in a little more detail than in section 2.2.

The salt drop experiment and the corrosion of iron in anaerobic soils have been described. It should be clear that in aqueous solutions that have been de-oxygenated, the only cathodic reaction possible is hydrogen evolution and that in neutral solutions this occurs at a very low rate. The two electrode reactions produce Fe^{2+} and OH^- ions. When the solubility product of $Fe(OH)_2$ is exceeded, precipitation occurs. The only other reaction is an extremely slow chemical reaction between iron and water, which produces Fe_3O_4. Neither reaction results in compounds that give protection. Nor is this surprising. The protective film on iron is γ-Fe_2O_3 which only forms at comparatively noble potential values (cf. section 2.8). Its formation reduces the anodic dissolution rate. Some inhibitors promote the growth of γ-Fe_2O_3 and in discussing ferrous materials they are described as anodic inhibitors to distinguish them from other inhibitors.

Anodic inhibitors include oxidizing anions, chromates and nitrites, and some non-oxidizing anions containing oxygen: phosphates, molybdates, tungstates, silicates, benzoates, etc. These function in solutions that are neutral or alkaline. In acid solution γ-Fe_2O_3 is unstable (cf. the Pourbaix diagram for iron) and dissolves by the mechanism of reductive dissolution.

When an oxidizing agent is added to a de-aerated solution, γ-Fe_2O_3 is formed:

$$Fe + NaNO_2 + H_2O \rightarrow \gamma\text{-}Fe_2O_3 + NaOH + NH_3$$

$$Fe + Na_2CrO_4 + H_2O \rightarrow \gamma\text{-}Fe_2O_3 + Cr_2O_3 + NaOH.$$

In the first case, ammonia, which is soluble, helps slightly to maintain alkaline conditions, in addition to the sodium hydroxide. In the second case, chromic oxide is incorporated into the growing film.

Upon the formation of γ-Fe_2O_3, which thickens logarithmically, the low corrosion rate in the de-aerated solution is reduced to an extremely low value. The effect of film formation is equivalent to inserting a high resistance in the anode reaction circuit. It is therefore highly polarized and the mixed corrosion potential of the specimen under these conditions is shifted in the noble direction.

Only oxidizing agents are effective in inhibiting the corrosion of iron in de-aerated neutral waters. The other anions, that are not oxidizing, require the presence of oxygen in the water. Oxidizing agents also inhibit in the presence of oxygen which lowers the minimum effective concentration of inhibitor that is required.

It is important to note at this point the protective role that oxygen can play in promoting the growth of protective films. In previous sections, e.g. 2.2, oxygen has been described as a cathodic depolarizer, but it can also function as an inhibitor, if it reacts with iron chemically to form γ-Fe_2O_3, with Fe_3O_4 as an intermediate product. The two oxides are mutually soluble in all proportions, so that the transition from one to the other can occur gradually. The concentration of oxygen needs to be high since it is being reduced as well as forming the protective film at the same time, although, as the film forms, the cathodic reaction rate will be diminished. The dual role of oxygen was demonstrated very simply many years ago when it was found that iron specimens that corroded in solutions containing air remained bright if rotated vigorously in the same solution[81]. Weak points in the film of Fe_3O_4, that were initiation sites for corrosion under stagnant conditions, were strongly oxidized under stirring conditions, where the effective concentration of oxygen was raised and the protective film therefore plugged the weak points. The film of Fe_3O_4 would also be gradually converted to γ-Fe_2O_3. A similar effect was observed for zinc specimens. Increasing the oxygen pressure above a solution also promotes the onset of passivity.

The double function of oxygen in corrosion reactions was shown up more fully by experiments, in which iron was exposed to drops of distilled water, in atmospheres containing varying compositions of oxygen and nitrogen[82, 83]. The results are illustrated in Fig. 66. The greater the proportion of oxygen, the greater was the probability that oxygen would be protective, but also the greater was the amount of corrosion that occurred if the oxygen failed to protect. This dual function of oxygen is a most important feature of corrosion reactions

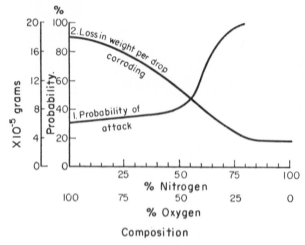

FIG. 66. Inception probability and conditional velocity of the corrosion of iron below drops of distilled water in oxygen/nitrogen mixtures[82,83].

of iron in neutral solutions. In strongly acidic solutions γ-Fe_2O_3 will not form and oxygen can only serve as a cathodic depolariser. Furthermore, under these conditions where the film cannot form, oxidizing agents, that would otherwise inhibit, will also provide alternative cathodic reactants and thereby increase the corrosion rate. Thus both pH and oxygen content are important variables in determining the effectiveness of anions in inhibiting the corrosion of iron.

When a de-aerated solution of iron that is not inhibited is saturated with air, the corrosion rate will increase very considerably, by as much as one or two orders of magnitude[84], because of the extra cathodic reactant provided by the oxygen. The oxygen also tends to form hydrated oxides containing ferric ions. These are highly insoluble and their precipitation lowers the equilibrium concentration and also therefore the activity of ferrous ions in the solution. The potential of the anodic reaction is therefore shifted in the active direction, a situation that is shown in Fig. 67 and which can be readily understood by reference to the Nernst equation. It is worth noting that the mixed corrosion potential under de-aerated conditions, E_1, is not necessarily

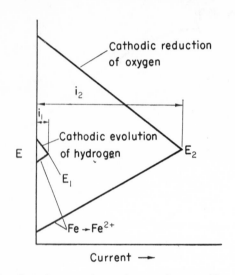

FIG. 67. The effect of oxygenation on the corrosion rate of iron in de-oxygenated neutral solutions. The corrosion rate is increased from i_1 to i_2.

lower than the mixed corrosion potential under air-saturated conditions, E_2. The difference between them gives no indication of the difference in the two corrosion rates. Interpretation of changes of corrosion potential (cf. section 2.9) can only safely be made when it is quite clear that the polarization characteristics of one of the two electrode reactions remain constant.

Oxidizing inhibitors function at low concentrations (10^{-3} M), but non-oxidizing inhibitors are only effective at much higher concentrations, since part of their function is to buffer the solution on the alkaline side of neutrality. These inhibitors interfere with the anodic reaction by plugging the film of Fe_3O_4 and converting it to γ-Fe_2O_3. In the case of chromate inhibition, the uptake of chromium into the growing film, which follows a logarithmic law, has been studied by analytical and radiotracer techniques[85]. If the iron is left exposed to air for long periods prior to immersion in water containing chromate ions, the amount of chromium entering the film is lower than if the iron is immersed immediately after abrasion. The uptake is also less in water containing oxygen, as compared with water that is oxygen free. These results reasonably suggest that the film is repaired by oxygen and that the Cr_2O_3 repairs such gaps as are missed. Oxidizing inhibitors function without oxygen and it is possible to explain this in terms of the chromate 'supplying its own oxygen', since $2\,CrO_4^=$ becomes Cr_2O_3, i.e. after the reduction of the chromium valency it is combined with fewer oxygen atoms than before.

Since oxygen itself acts as an inhibitor and is clearly necessary for the formation of γ-Fe_2O_3 from any lower compound, it must be present if non-oxidizing inhibitors are used, since these do not 'supply their own oxygen'. Much higher concentrations of this type of in-

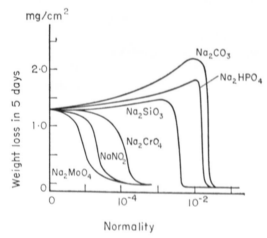

FIG. 68. Weight loss vs. concentration curves for several oxidizing and non-oxidizing inhibitors[86].

hibitor are therefore required than for the oxidizing type. This is illustrated in Fig. 68, where the two classes are clearly distinguishable from each other in this respect[86].

It is important to note from Fig. 68 that insufficient anodic inhibitor can actually increase the amount of corrosion. If most of the gaps in a film are plugged, then the attack that does occur will be concentrated upon the unplugged sites. Since corrosion reactions are often under cathodic control, the larger cathode area (which will include all the plugged sites) therefore raises the anodic dissolution rate and deep penetrating attack results. The exception here is inhibition by benzoates. If insufficient is added the attack is widespread and uniform. This observation helps to explain the inhibitive mechanism of the benzoate ion. Such a large molecule is unlikely to be built into the protective oxide. Instead, it is possible that the ion is continually adsorbing on to and desorbing from anodic areas. Its effect lies in shielding a large proportion of the exposed surface at any one moment from the solution, so that the exposed part is then oxidized to γ-Fe_2O_3.

The ability of benzoate ions to adsorb on an iron specimen is related to its potential. Sodium benzoate only protects iron specimens whose potential is more noble than -280 mV E_H scale[87]. Iron coupled to zinc, or already rusting, is not protected by benzoate ions. The case of their adsorption appears to be related to the electrocapillary maximum of the metal, a subject on which there is little known for solid surfaces.

One explanation of the mechanism of inhibition is that the oxidizing inhibitors are so readily reduced that large local cathodic currents cause the adjacent anodic areas to become passive[88]. The oxidizing anodic inhibitor is therefore sometimes called a passivator. The description agrees with the observed shift of potential. With non-oxidizing inhibitors such a mechanism is less likely. A film does form which contains insoluble 'plugs', e.g. of ferric phosphate in the case of inhibition in phosphate solution.

The mechanism of inhibition is still not clear. Film growth may in some cases be the result of inhibition rather than its cause, since sometimes, under unusual conditions, films continue to grow after inhibition has been established until they exhibit interference colours. As in the case of passivity, aggressive anions, e.g. Cl^-, make inhibition more difficult to obtain and their presence makes a higher concentration of inhibitor necessary than in their absence.

It is also possible to obtain cathodic inhibitors. Anything that interferes with the cathodic reduction reaction is classified under this heading. In closed water systems, for example, additions are made which react with the limited amount of oxygen present, e.g. hydrazine in boiler waters, and these scavengers are effectively cathodic inhibitors.

Magnesium and calcium salts, which occur widely in natural waters, are also cathodic inhibitors. Their hydroxides, which have low solubility products, precipitate in cathodic regions and the film that is thrown down on the metal surface retards oxygen reduction since electrons have to pass through it. Such scaling occurs very readily in hard water. A scale reduces the initial corrosion rate considerably, although it is obviously a nuisance if, for example, the cross-section of the pipe is reduced significantly or if desirable heat transfer properties are interfered with by its formation. Since anodic and cathodic sites occur adjacently, the products of both reactions interact and scales usually consist of a mixture of rust formed from the metal, with hydroxides and carbonates formed from the water.

Some inhibitors interfere with both the anodic and cathodic reactions. Such mixed inhibitors include the commercially available polyphosphates (sometimes called condensed phosphates). These form a highly protective deposit when used with steel. Their action is enhanced by the presence of calcium ions. The mechanism of protection is little known and a rather unusual feature, illustrated in Fig. 69[84], is the optimum concentration required for maximum protection. Both below and above it the corrosion rate increases. The increase at higher concentrations is caused by the complexing action of the phosphate[84]. Polyphosphates, which require oxygen in the water and which buffer on the alkaline side, are unsuitable for use in hot water since they undergo hydrolysis. Phosphates and silicates are used instead.

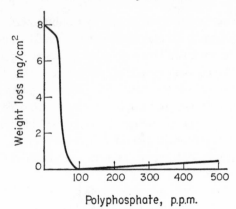

FIG. 69. Weight loss vs. concentration curve for iron in polyphosphate
solution[84].

Although the formation of a scale helps to reduce the corrosion of
metal pipes, only a thin layer is desirable and the aim in water treatment is to eliminate scale growth. Unless this is done, the working
diameter of the pipe will gradually diminish and at some stage a
descaling operation will have to be undertaken. Polyphosphates are
useful in this respect too, since they reduce the precipitation of $CaCO_3$.

The carbon dioxide content of water is very important, not only in
terms of its concentration but also in the form in which it is present.
There are four possible forms:

(1) Carbon dioxide in the combined form as carbonates.
(2) Carbon dioxide in the combined form as bicarbonates.
(3) Carbon dioxide in the water, which maintains the carbonate/bi-
carbonate mixture with which it is in equilibrium.
(4) Free carbon dioxide. This is aggressive, unlike the other three
forms.

Up to a pH of approximately 8·5, free carbon dioxide and bicarbonate
can exist in equilibrium, while at higher pH values the bicarbonate ion
is replaced by the carbonate ion[89].

In hard waters carbon dioxide will actively participate in scale
formation, while in soft waters, which are poor scale formers, the

carbonic acid attacks the metal. This may produce dissolution, e.g. lead, which is a toxic hazard. Copper is also susceptible to rapid corrosion in aggressive soft waters.

It is clearly important to know whether water contains aggressive carbon dioxide. One test consists essentially of passing the water over chalk chips and measuring its pH both before and after. If it increases, then the water is aggressive and it will prevent the establishment of a film, while if it decreases, then the water is film-forming. The change is referred to as the Langellier Index.

Many soft acidic natural waters are hardened by additions of lime and then supplied at pH 7–8. Film-forming properties vary considerably. Chloride ions tend to hinder film formation. Nitrate ions are also harmful, though usually present in much smaller amounts. Sulphates, which are subject to bacterial degradation, attack concrete and can interfere with inhibitors. Silica is a constituent of natural water and is comparatively harmless. It is not a substitute for silicate addition. Organic matter can cause severe pitting of metal surfaces where it settles if it excludes oxygen. It can also cause overheating because of its poor thermal conductivity. Oil on water can promote bacterial activity by cutting off the oxygen supply and it may of course contain aggressive constituents that dissolve in the water. Not all bacteria are harmful. Some, for example, found in parts of England, exert a strong inhibitive action upon the corrosion of copper. Organic acids from moors and peats make soft waters extremely aggressive to steel.

These are merely some of the variables and the number is considerable. Furthermore, small alterations can produce large effects. The most important and general variables are carbon dioxide and oxygen content, pH, hardness value and chloride content.

The importance of the conductivity of the electrolyte in electrochemical corrosion has been emphasized in section 2.2. In supply waters the total concentration of the dissolved ionized constituents (referred to in the United Kingdom as Total Dissolved Solids—TDS) is a factor in determining their corrosiveness. This matter can be particularly important in waters used for cooling. Evaporation losses and automatic make-up can produce large increases in TDS and thereby result in an untenable rate of corrosion. Such problems are avoided by diluting the remaining water by discharging from it a volume equal to the evaporation loss and then filling up with new supply water, a procedure referred to as 'bleed-off'.

So far only ferrous materials have been considered. Other metals also need inhibitive protection. In many cases the same inhibitors are found to be effective. Chromates, silicates and polyphosphates are effective for zinc and the first two are used for aluminium. Tin plate is often given a chromate dip treatment as the last stage in processing. Chromate treatment of waters is usually good for systems employing several metals together. For other metals highly specific inhibitors are used. Fluoride ions inhibit the corrosion of magnesium and sodium mercaptobenzothiazole that of copper. The latter is used, for example, in some antifreezes with a borate buffer. It is also used as a vapour phase inhibitor for copper which resists tarnishing at ambient temperatures in somewhat aggressive humid conditions when wrapped in paper impregnated with it. There are many uses for vapour phase inhibitors in storage or for temporary protection, etc. They are often used either coated or impregnated into a packing material. Certain amines or organic esters are used, e.g. dicyclohexyl-ammonium nitrite, which is very effective on steel. Aluminium is sometimes wrapped in paper impregnated with chromate. Moisture in the paper and atmosphere help form a very thin aqueous chromate layer on the metal surface. This is therefore not vapour phase inhibition. There is much variation possible in this type of treatment. The corrosion resistance of aluminium, for example, has been increased by autoclaving specimens with 1–2 wt% aqueous chromic acid at pH 1 at 160–180 °C for 20–50 hr[90]. Other chemical surface treatments are described in section 3.6, since they are often applied as paints.

Octadecylamine and hexadecylamine are sometimes added to boiler waters. They are surface-active compounds which form a water-repellent film on the inside surface. Vapour phase inhibitors have also been used to protect idle boilers, e.g. cyclohexyl-ammonium carbonate.

Inhibition is a complicated subject and can only be successfully applied in difficult circumstances by a person able to draw from a wide

field of knowledge. There are many potential hazards. Nitrites should
not be used with brass since the decomposition product, ammonia, can
cause stress corrosion cracking (cf. section 4.3). Aluminium oxide
dissolves in both acids and alkalis and pH control is of major importance, although strong oxidizing conditions may stabilize the oxide film
outside the normal pH limits, e.g. hydrogen peroxide added to weakly
alkaline solutions or concentrated nitric acid.

In the practical sense, anions are used for anodic inhibition, but it
must not be supposed that only anions function as anodic inhibitors.
For stainless steels, for example, the redox system Fe^{2+}/Fe^{3+} can
promote inhibition through passivity. At low concentrations and at
active potential values the reduction serves as an additional cathodic
reaction and increases the dissolution rate. But, as in the examples

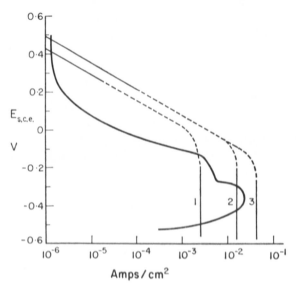

Fig. 70. Superposition of reduction curves for ferric ion and of the
anodic polarization curve of type 410 stainless steel in N H_2SO_4. (1)
0·010 M Fe^{3+} at 55 cm/sec; (2) 0·066 M Fe^{3+} at 55 cm/sec; (3) 0·066 M
Fe^{3+} at 100 cm/sec. In case (3) the limiting diffusion current density for
the ferric ion is greater than the critical current density for the steel and
the ferric ion therefore passivates the steel[91].

quoted in section 2.8, if the cathodic current density is increased beyond the critical current density of the anode reaction, then the metal becomes passivated. The situation is shown diagrammatically in Fig. 70, which illustrates the influence of both inhibitor concentration and flow velocity upon the corrosion of a ferritic stainless steel in the presence of ferric sulphate[91]. This type of inhibition which induces passivity is somewhat different from inhibition by chromates and nitrites, since the latter lose oxygen as part of the reduction process. Since these ihibitors are called passivators by some authorities, the term must also include not only redox systems like Fe^{2+}/Fe^{3+} in the example just cited, but also the system H_2/H^+ on stainless steels containing noble alloying elements as described in section 2.8. The nomenclature is a little untidy, but it should not be allowed to obscure the explanation of the results that has been given.

Acid inhibitors, often referred to as pickling inhibitors or restrainers, are used when acids are likely to attack metal. The common uses occur in removing scale from metal during hot fabrication and for descaling water pipes. In the first case very thick scales (*ca.* 0.1 in) form during rolling. They must be removed quickly and economically. Sulphuric and hydrochloric acids are used. Inhibitors reduce the loss of metal, the loss of acid, and the unpleasantness of working conditions.

Acid inhibitors are mainly organic molecules containing nitrogen and sulphur in polar groups. They adsorb on to the bare surface of the metal and shield it from further attack by the acid. They do not prevent attack by oxidizing agents that are present, e.g. ferric ions derived from chemical dissolution of magnetite, nor do they reduce the rate of pickling where it occurs. Some actually increase this rate. They are in dynamic equilibrium with the surface and they are not consumed directly, like anodic inhibitors in neutral solutions. During treatment of metal parts there is a gradual reduction of inhibitor concentration, however, since inhibitors are removed, adsorbed on pickled surfaces.

Acid inhibitors interfere with both the anodic and cathodic reactions to different degrees, dependent upon the type of molecule being used. Quinolines and substituted quinolines are mainly anodic inhibitors, but

at high concentrations they are cathodic inhibitors as well[92]. Thioureas and substituted thioureas inhibit both reactions, being more
cathodic in their inhibitive effect at low concentrations.

Only over narrow limits of concentration do these inhibitors adsorb
in a monolayer and obey Langmuir's isotherm. Generally the behaviour has not been found to be systematic. Sometimes at low concentrations each molecule may protect an area around it, the size and
shape of which will depend upon the molecule structure, so that protection, as measured by the decrease in corrosion, increases by a
power a little greater than unity as the concentration is raised. Sometimes the power is less than unity, suggesting that the molecules are
forming a second layer before the first is complete.

In Fig. 71 the movements in corrosion potential against corrosion
rate are shown for a number of inhibitors[92]. A shift of potential in
the noble direction shows that the inhibitor is anodic, a shift in the
active direction that it is cathodic. All the inhibitors are predominantly

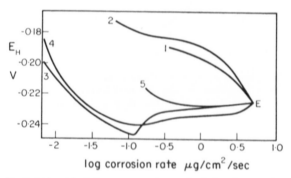

Fig. 71. Relationship between corrosion potential and corrosion rate
at 70 °C for several inhibitors[92]. 1, quinoline; 2, 2:6-dimethylquinoline;
3, *o*-tolylthiourea; 4, *m*-tolylthiourea; 5, thiourea.

anodic at high concentrations, where a surface is practically completely
covered with inhibitor molecules. This is likely, since any small gaps
that are left are more likely to allow a hydrogen ion to reach the
surface and be discharged than permit the passage outwards of a
comparatively large cation.

Both the size and shape of the inhibitor molecule affect its inhibitive powers, as is shown in Fig. 72, where substitution of different radicals can be seen to have a marked effect[91]. Steric hindrance factors have been little investigated but are clearly of some significance. The potential of the specimen is also of importance. Unlike inhibition by benzoates, where the value of the potential is also important, slight cathodic polarization often improves the inhibitive powers of organic molecules considerably.

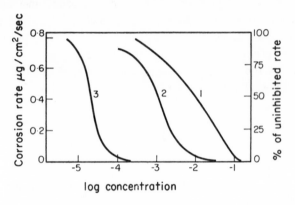

FIG. 72. Influence of inhibitor concentration, c moles/l, on corrosion rate at 70 °C for several inhibitors. Uninhibited rate $= 0.84 \mu g/cm^2/sec$[92]. 1, quinoline; 2, N-ethylquinoline; 3, 1-tolylthiourea.

Organic molecules can affect inhibitive properties in neutral solutions. Such molecules include tannins, dextrin, gelatin, agar, some oils and miscellaneous breakdown products rather than the polar molecules used in acid inhibition. These have not been investigated mechanistically to any large extent. Gelatin appears to be a cathodic inhibitor at pH < 4.7[93], a value that probably represents the iso-electric point. Usually their discovery is largely empirical and their use pragmatic.

Nearly all acid inhibition work has been concerned with ferrous materials. Aniline and phenol derivatives have however been found to protect titanium in strong hydrochloric acid solution[94]. In low concentrations the inhibitors were found to promote corrosion, which is an unusual feature, suggesting that a different mechanism is at work from that during ferrous inhibition.

Metals with high hydrogen overpotentials are successfully used to reduce acid attack, e.g. antimony, tin and arsenic. They 'plate out' on to the bare metal and polarize the hydrogen evolution reaction. Thus acidified antimony chloride is used for cleaning metal films from steel and for derusting solutions, sometimes with stannous chloride added.

3.5. METAL COATINGS

Metals are coated with metals in several ways and for each method there is frequently a considerable variety of techniques involved. Steel is treated in all possible ways. For example, steel parts are often dipped into molten zinc or tin, or heated with zinc dust (Sherardizing). Both coating metals can be sprayed on, while the modern method of applying tin is by electroplating. Vapour deposition is sometimes used, where the metal forms a convenient compound that is both volatile and readily decomposable. Where the temperature is high, the coating metal will diffuse into the base metal and this may result in intermediate phases being formed. It is difficult to produce a perfect continuous coating of metal in electrode deposition, since (a) hydrogen evolution often causes tiny cracks in the deposit, (b) high compressive and tensile stresses arise in deposits, which may also cause tiny cracks to form and (c) the adhesive properties of the underlying metal surface vary considerably if residual grease or etching reagents are left on. Some chemical deposition processes deposit non-metallic material with the metal and produce a layer with a pore structure. All that can be said is that each method of application gives deposits with different characteristics. Furthermore, many are porous (or at least discontinuous). Since it is commonly not possible to produce perfect coatings, some have been developed in which the density of cracks has been deliberately increased in coatings for which the underlying metal is the anode. Since there is only a certain total cathode current available, the creation of many anodes is intended to ensure that the amount of corrosion occurring at any one point is small in comparison with that occurring at an anode in a coating containing few cracks.

The shape, size and distribution of coating discontinuity varies considerably, depending upon the metal and method employed. The corrosion behaviour of a coating will depend upon these variables. The coating itself constitutes a galvanic couple, although only a small proportion of the underlying metal may be exposed, e.g. 10^{-6}.

Metal coatings are often applied in order to improve the corrosion life of the underlying metal and the several mechanisms by which this improvement is achieved will be described. Coatings are also applied for several other reasons, e.g. (a) for increased wear resistance (chromium), (b) for consistent low electrical contact resistance (preferably gold, often silver), (c) for high and constant reflectivity (chromium), (d) oxidation resistance (aluminium on iron), to mention only a few.

A large quantity of mild steel is covered with zinc (galvanized). Zinc is anodic to iron and when moisture penetrates to the base metal the zinc is sacrificially corroded. Its protection ceases when such an area of steel is exposed that the polarization caused by the zinc is rendered insufficient, usually at the centre of the bare steel. The amount of zinc in the layer is the most important protective factor. The thickness of deposit governs its capacity for protection, regardless of the method of deposition, although there are minor variations. The protective properties of zinc on iron lie not only in the sacrificial action, which is clearly limited, but also in the blockage of corrosion sites which occurs when the zinc ions combine with the hydroxyl ions produced by the cathodic reaction on the iron surface and precipitate as the hydroxide within the minute gaps in the coating. Galvanized surfaces are often subsequently chromated, as described in section 3.6.

Zinc coatings are widely used for protecting steel from aqueous environments. In the presence of oxygen, zinc hydroxide is precipitated from the products of the anodic and cathodic reactions and it acts as a barrier for the oxygen reduction reaction. Carbon dioxide in the water will react with the hydroxide to form somewhat more soluble carbonates and will thereby raise the corrosion rate. Soft waters therefore require a thicker coating of zinc than hard waters for the same protection, because their film-forming action is much weaker. As the temperature is raised, the corrosion rate increases, but above 60 °C the corrosion products change from a loose gelatinous form to a very condensed form and the corrosion rate is much reduced. Sometimes reversal of the galvanic potential difference occurs and attack on any exposed parts of the iron is then considerably stimulated.

Coated steel in stirred solutions can only be safely used in the pH range 6–12·5.

Cadmium plating is also sacrificial to iron, but the galvanic potential difference developed is less than with zinc. Cadmium appears to withstand marine environments better than zinc; its chloride is less soluble and therefore probably more protective. It withstands industrial atmospheres less well than zinc; its sulphate is indeed more soluble and sulphates form a major part of corrosion products in these atmospheres (cf. section 2.7). Cadmium coatings are generally superior to zinc coatings under humid indoor conditions; they corrode parabolically indoors while zinc coatings corrode rectilinearly.

Tin plate represents a different corrosion situation, since tin is cathodic to iron under normal conditions and might be expected to accelerate corrosion at breaks in the coating by providing a large efficient cathode area. This can happen and pin hole corrosion is found to occur on tin plate, but it is not common.

Very large quantities of tin plate having only a thin layer of tin are used for containing food. Thicker coatings are used for milk churns, petrol tanks, etc. Tin is fairly resistant to corrosion and stands up well to foodstuffs. Many organic acids that are present in foodstuffs or in the fruit juices contained in cans form complexes with tin. The activity of the stannous ions is therefore lowered and the tin becomes anodic with respect to iron in the specific environment. The resistance of tin is also partly due to its high hydrogen overpotential. In the absence of oxygen, the only other possible cathodic reaction proceeds at a very low rate. This is absolutely necessary since the evolution of hydrogen within a closed tin can is a danger that must be avoided. Even if breaks do occur in a tin coating the iron, which is protected, will also evolve hydrogen very slowly, since it too has a high hydrogen overpotential. Inhibitors are sometimes added while sometimes they occur naturally in the contents which may, of course, also contain accelerators.

On the outside of a tin can conditions are different and pin hole attack if it occurs will start from this side. Usually tin plate is given a chromate dip before being fabricated into containers. This affords additional protection. The inside is often lacquered also. If labels are used, they should be stuck on with non-corrosive glues.

The corrosion of tin cans is a complex process, since there are numerous factors which become important under different conditions. Thus sulphur compounds react with tin and produce films that interfere with its protective action. An important feature is the formation of an iron–tin compound $FeSn_2$ during 'flow-brightening' of electrolytic tin plate, or during hot dipping. This is inert under the conditions existing inside a tin can. Stannous ions in solution tend to slow down the dissolution of steel by affecting a degree of anodic inhibition. There are other important variables too. To cope with them all, a series of tests has been devised which relates the long life behaviour of a tin can (sometimes called its shelf life) to its contents.

Atmospheric corrosion of tin plate arises from the porous nature of thin layers of tin. Rusting will start within the pores, although under alternate wetting and drying conditions effective plugs may be formed.

Tin is also used as a coating for copper and brass since it provides a non-staining protective layer for handling foodstuffs, water supplies, cooking vessels and helps to avoid catalytic oxidation of products by copper. It also provides a good soldering surface.

Terne plate is steel coated with lead–tin alloys containing 5–25% tin. They are sprayed or electrodeposited. They have very good resistance to atmospheres containing sulphurous compounds.

Tin–zinc coatings, which are electrodeposited, suffer from dezincification, but they are good for tool components, aircraft undercarriages and for reducing galvanic attack on steels in contact with aluminium alloys.

Tin–nickel coatings containing 65% tin have good resistance to atmospheric attack, including those containing sulphurous compounds. In aqueous solutions they are passive and withstand vinegar, alkalis, fruit juices, etc. The coatings accelerate attack on the underlying metal, but this can be avoided by applying the coatings very carefully in the first place and by an intermediate thin layer of copper. Tin–nickel coatings are extensively used, largely for indoor purposes.

Lead is used mainly on account of its resistance to attack by a wide range of chemicals. Since its mechanical properties are poor, it is used as a liner for acid baths, etc. It has very good resistance to atmospheric corrosion.

Copper is largely electrodeposited as a coating. It is generally covered with nickel or zinc and its corrosion resistance must be considered in conjunction with these.

Copper improves the resistance of nickel plate to the atmosphere, particularly if the nickel is chromium plated.

Nickel coatings, which are electrodeposited or chemically deposited, are used widely to protect many metals, particularly steel. Nickel 'fogs' in industrial atmospheres. It catalyses the conversion of sulphur dioxide to the trioxide and this produces nickel sulphate as the final product. This catalytic reaction can be poisoned by prior exposure of the clean nickel to hydrogen sulphide gas. Nickel is commonly plated with chromium.

Nickel is generally cathodic to steel and its use is largely based upon the inherent resistance of nickel metal. It must be a continuous coating. The most important parameter is the thickness of the coating applied. This coating inevitably has pores and these lead to pitting. Nickel corrodes preferentially to chromium and when attack reaches the steel base undermining of the nickel deposit is initiated. Industrial atmospheres containing sulphur are extremely aggressive towards nickel plate.

Chromium is a reactive metal, but its highly resistant film of Cr_2O_3 reduces attack to a negligible amount and makes it a widely used coating which is normally electrodeposited, although vapour deposition from the iodides and chromizing are both used.

The electrodeposited layer is very porous and contains many cracks, so that protection is most economically and usefully obtained by using nickel underneath. Crack-free deposits are possible and exhibit better corrosion resistance.

The absence of oxygen and the presence of aggressive anions, e.g. Cl^-, can rapidly cause serious pitting on chromium plate. Green nickel salts become visible as stains on the plate and these are followed at a later stage by rust.

Noble metals are very limited in their use as coatings by their cost. They are generally used where their specific properties demand it; this is usually on a very small scale. Intermediate undercoats are frequently used in order to reduce the large galvanic potentials that might otherwise develop. In addition, because the platinum metals are so expensive, only a very thin deposit is ever used and since it may be porous it requires a protective underlay.

In electrodeposition the variation in the coating thickness is an important factor. Each bath that is used has a throwing power, which is a measure of the ability of a solution to deposit an even coating. It usually varies with the cathode current density and is lowered by raising the temperature and stirring the solution. If the throwing power is very poor, a recessed corner of an object may be left with no deposit at all. The throwing power is governed by those electrochemical factors that determine the distribution of equipotential lines. Consequently, the term is used about any electrochemical situation and it is not restricted to one depositing metal.

3.6. CHEMICAL COATINGS

Chemical coatings include thick organic layers of plastic and rubber, cements and glasses and ceramics, but these will not be considered in this book. Instead chemical coatings will include only those coatings produced by a chemical or electrochemical reaction that the metal takes part in.

Electrochemical coatings are produced by anodizing. In this process a coating, usually an oxide, is produced on a metal surface by making the metal an anode in a cell. Aluminium is the metal most widely treated in this way, but other metals exhibit similar anodic oxide growth, e.g. titanium, tantalum.

Anodic treatment of aluminium in certain baths, e.g. 5–10%

sulphuric acid, 3–10% chromic acid, 2–5% oxalic acid, produces a film of alumina (which is an oxide that exhibits several different hydrated forms) that has a pore structure superimposed on a compact layer, as illustrated in Fig. 73. This structure only occurs in solutions in which the oxide has a moderate solubility. If the solubility is very low (e.g. as in boric acid), only a compact thin oxide is produced. The pores can be closed by exposure to steam and dyes can be sealed in them, a process that results in the wide decorative use of anodized aluminium.

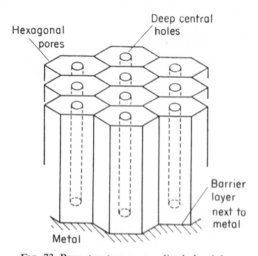

FIG. 73. Pore structure on anodized aluminium.

Sealed anodized aluminium oxides have a very good corrosion resistance, the more so if dichromate ions are sealed in them, since these exert an inhibitive effect. Much depends upon the pore size, which is determined by the electrolyte and the forming voltage employed, since pores become more difficult to seal completely as they become larger. The thicker the film, the greater the protection afforded and in Table XI the minimum thickness of anodic coatings for different atmospheres, as laid down by a British Standards Specification, is shown.

TABLE XI. MINIMUM THICKNESS OF
ANODIZED COATING
(B.S. 1615/1958)

	Thickness
Permanent outdoor exposure, minimum maintenance	25 μ
Permanent outdoor exposure, well maintained	15 μ
Indoors	5 μ

The film is largely composed of alumina monohydrate which is a comparatively inert substance. Any weak points arising, for example, from impurities may suffer very severe attack. The purity of the aluminium is all-important. Sometimes containers are left not anodized so that attack will, if anything, be widespread and the risk of perforation is thereby avoided.

Magnesium fluoride coatings are also formed anodically. They are used mainly as a base for painting.

Chemical coatings are very widely used for providing a rapid easy protective coating which can be primed and painted.

Phosphate coatings are widely used for protecting steel. In addition to their inhibitive properties, they provide a good surface for subsequent painting. A large proportion of the world's steel products get a phosphate treatment at some stage. Other metals are phosphated too, except for aluminium which is usually treated with chromate solution.

Phosphate baths contain mixtures of iron, manganese and zinc primary phosphates and dilute phosphoric acid. The latter decomposes in dissolving iron and this reaction eventually leads to the deposition of insoluble tertiary phosphates and a small proportion of iron phosphates on the surface of the iron. Cathodic treatments accelerate the process and so do both oxidizing and reducing agents.

There are many variables in phosphate treatments which are primarily designed to produce coatings of different thicknesses and grain size. The weight of coating usually varies from 50 to 300 mg/ft². The quickly forming thin coatings are used where painting is to follow, while thick coatings are used where oily conditions and heavy friction are likely to be met.

Careful compositional control of baths must be kept, since low pH will lead to etching and high pH to sludge formation. Residual grease on the metal has a pronounced deleterious effect upon the process. Once formed, however, the coatings readily absorb grease and this generally increases their corrosion resistance.

Chromates are also applied as a chemical coating to many metals. On ferrous materials no direct coating process is available and a chromate dip is only given after a phosphate treatment. Activators, e.g. Cl⁻, are used to eliminate passivity of the metal surface which would otherwise prevent chromate film build-up. The bath contains a hexavalent chromium compound and a mineral acid and the chromium suffers some reduction from the hydrogen that is evolved. The film consists of a mixture of chromic oxide, chromium trioxide and the oxide of the metal. The pH of treatment depends upon the stability of the metal oxide, which must be removed.

Aluminium, zinc and cadmium are all treated by various chromate baths. Copper can be brightened by dichromate dips, but the corrosion resistance is only slightly improved. Magnesium is given a chromate dip before painting. The tarnishing resistance of silver is increased by treatment with chromate. Tin plate is usually treated with chromate.

Chromate coatings are used either alone or followed by the application of paint. They cease to be effective once moisture has blacked out all the hexavalent chromium or if they are heated. They can be coloured or lacquered.

3.7. PAINTING

Metals are painted as a protective measure against corrosion and in order to make them visually attractive. The initial step of surface preparation is most important. This includes not merely degreasing the surface but also ensuring that it is dry and free from scale. Degreasing can be done with organic solvents (vapour degreasing baths are very common), but it is often more effective and cheaper to clean with an alkaline liquid. Much depends upon the nature of the original grease and other surface contaminants. Commercial metal cleaners are usually alkaline, with pH values varying from 8 to 11·5. These are based on caustic soda, alkali silicate, phosphates, salts of higher fatty acids, etc. Modern synthetic detergents, which have excellent emulsifying powers, are also widely used.

Pickling varies considerably in practice. Anodic and cathodic pickling, ultrasonic cleaning, fused salt baths and many different kinds of mechanical methods are used. The cheapest, quickest and most efficient method is usually aimed at, but is often only achieved after much experience.

Drying a metal surface is not always easy, particularly for outside structures in humid atmospheres that are to be painted. Scale removal is also difficult and of considerable importance. Steel that has been hot rolled, etc., nearly always has residual millscale which is very firmly attached, even though the surface may have been pickled near the end of processing. This will absorb moisture and produce a general lowering of adhesion of the painted surfaces, which may also blister if the scale undergoes reactions with water that cause a volume expansion. Furthermore, the scale on steel consists of oxides that exhibit some electronic conduction and these therefore behave as fairly efficient cathodes, able to stimulate the exposed parts of the surface. Local cells may be established where the moisture is absorbed and pitting will then start. Extreme care must be exercised in removing scale, even though the cost of manpower involved is very considerable. Sand blasting in air or inhibited water is the most common method employed, although flame cleaning is also used, in which after a degreasing operation the scale is heated in order to expand it rapidly so that it will flake off while the underlying metal is kept comparatively cool. A protective coating is then quickly applied. Weathering is also commonly used where an unpainted structure is left exposed to the atmosphere for as long as six months. The millscale undergoes dimensional changes and flakes off Subsequent physical removal is then very much easier. Great emphasis is laid upon the complete removal of scale. It is the most important step in painting, in so far as good preparation and poor paint is better than bad preparation and good paint.

Once the metal surface is cleaned it can be painted, a process that is done in several stages. Firstly the surface is given a protective wash, commonly a phosphate bath, as already mentioned in section 3.6. This prevents the rusting of the descaled surface and needs to be done fairly quickly after the removal of the scale. Then it is primed and finally it is painted. This is the general sequence. In practice, each step may require several operations, or two steps may be combined in one application.

The paint consists of a drying oil, called the vehicle, that will form a film and a finely ground pigment. In addition, paints frequently contain thinners, which lower the viscosity for ease of application, driers, which accelerate the oxidation of the vehicle from a liquid to a hard skin, extenders, which aid the dispersion of the pigment, and sometimes plasticizers, which reduce the brittleness of the film. Within that general description, much variation is found and thousands of different paints are made. Many use linseed oil as the vehicle. Some contain natural varnish. Many more are made from the multiplicity of organic materials that are commonly available, often in a polymerized combined form, others require heating to hasten chemical and physical change, e.g. enamels, while lacquers depend upon evaporation and therefore require considerable care in mixing. Some paints are applied by electrophoresis.

With all the possible variety and treatment, there is much variation in the physical and chemical properties of films and therefore in the corrosion behaviour of painted metal parts. In the case of linseed oil paints it is important to understand that the paint film is not a continuous film in the sense that it does not cover the metal like a blanket. Apart from any flaws that it may contain, it is permeable to small amounts of oxygen, like a membrane. Furthermore, linseed oil decomposes slowly, particularly in sunlight, and forms a large number of organic acids. Some of these are aggressive, particularly to light metals, which therefore require thorough surface pre-treatment.

Since paints do not afford complete protection and since they may become aggressive during breakdown, priming paints are used. These also provide a very good surface for subsequent coats of paint. The primer material depends upon the metal that is to be protected and the environment that it is to withstand.

The most widely used primer is red lead. This consists of Pb_3O_4 containing small amounts of litharge, PbO, from which it is made by roasting. It has a characteristic reddish/orange appearance that can frequently be noticed on exposed metal structures, etc., during painting. Although it has been widely and continuously used for 2000 years, its ability to afford considerable protection is not fully understood. Undoubtedly it forms insoluble metal soaps (metal ion plus organic acid molecule) with the breakdown products of linseed oil and presumably in that way restricts their aggressive action. It also functions as an inhibitor, probably in combination with the acid radicals, since dilute solutions containing lead salts of these acids have been found to inhibit the corrosion of mild steel[95]. In the film, therefore, the presence together of both oxygen and lead soaps acts to inhibit the corrosion of the metal. The oxide is also mildly alkaline and will encourage protective film build-up in the absence of the decomposition acids. Red lead is aggressive towards many metals and its use is largely confined to ferrous materials.

Other lead compounds are also made up into primers. White lead, which consists of basic lead carbonate, and blue lead are both alkaline and protective for the same reason as red lead. Calcium plumbate, apart from being a useful primer, also has anti-blistering characteristics when used on metal parts immersed in sea water. Basic lead chromate combines the protection afforded by lead in paints with additional inhibition provided by chromate ions. Zinc chromate paints provide galvanic protection action by zinc, a mild alkalinity and chromate inhibition. It is used on light metals, where red lead cannot be used. In acid environments it may stimulate corrosion.

Extender pigments also may exert protection. Iron oxide is not inhibitive in itself, but helps to produce strong impervious films when used with chromates, red lead, etc. Iron oxide pigments, which are opaque to UV light, find much use in top coats as extenders, particularly in the micaceous form with which very good coverage is obtained.

Both zinc and aluminium metal are applied as paints. The zinc may be dissolved out eventually and the paint develops blisters in sea water. Although used by itself in large quantities, zinc is also mixed with other protective materials, to produce specialized protective coatings, e.g. zinc with sodium silicate, which will inhibit in the presence of oxygen and galvanically protect where inhibition is lost. Aluminium paint contains very thin flakes of metal which lie on top of each other, providing an inert covering. This will cathodically protect most metals if breaks occur in the oxide film on the aluminium. The paint has good moisture impermeability and is opaque to uv light. It is not used as a primer, but as a final extra protective and decorative coat.

Carbon black is used in paints as an extender in the outer decorative finishes. Should it be used in priming coats, then direct contact of carbon with the underlying metal can stimulate corrosion.

Copper paints are used largely for anti-fouling measures and they must be used with care, since they may stimulate galvanic attack, e.g. on aluminium.

Much work is done in mixing priming pigments in attempts to produce paints containing the optimum set of properties. There are of consequence many types of priming programmes, some of which are extremely costly to use, and many have been developed for highly specific circumstances.

A fairly recent surface treatment combines the initial surface coating operation with a priming coat. The mixture is referred to as a *wash primer* or *etch primer*. One composition (that of WP1) is given in Table XII. The zinc chromate in polyvinyl butyral is mixed with the acid just before use. It can be brushed or sprayed on and quickly followed by a further priming coat or final finish. Wash primers are used with steel, zinc, tin, magnesium, aluminium and other metals. They are in common widespread use.

The final coating is thinner than that obtained from the two separate treatments and may be less satisfactory in some instances. In many circumstances, however, a wash primer appears to be quite adequate.

TABLE XII. WP1 WASH PRIMER

Base material	
Polyvinyl butyral	7·2%
Zinc tetroxychromate	6·9%
Talc	1·1%
Isopropanol 99%	50·4%
Toluol	14·4%
Acid diluent	
Phosphoric acid 85%	3·4%
Water	3·4%
Isopropanol	13·2%

Apart from the chromate inhibition, the mechanism appears to depend upon the open structure of the zinc tetroxychromate lattice which is able to hold zinc oxide molecules. Zinc phosphate is then precipitated on the metal surface and stifles active sites. Although wash primers are used on both aluminium and magnesium and their alloys, these metals are both more often chemically treated and then primed and painted.

Aluminium and its alloys are usually treated with zinc chromates before painting. Crevices where oxygen may find it difficult to penetrate are particularly prone to corrode and care must be exercised in this respect. Chromate pastes are sometimes used as jointing compounds. Lead-based paints should be avoided. Magnesium is often treated with chromate and then a chromate primer followed by an aluminium primer and a decorative finish. The vehicle must be resistant to alkali, e.g. epoxyresin and not linseed oil.

All paint films provide considerable resistance to ionic movement and their ability to do so is a good measure of the protection that they provide to metal surfaces. Consequently the protection of a primer consists of several parts: (a) the inhibitive action of the primer pigment,

(b) the resistance of the film, and (c) the action of metal soaps, both in rendering films less permeable to water and electrolytes and in reducing the aggressiveness of the acidic film products. The ion exchange capacity of the film, which determines in part the passage of electrolytes, is an important chemical factor in paint technology[96].

3.8. CORROSION DESIGN

In this short outline of the methods of protection against corrosion the emphasis has been on the underlying principles and mechanisms. In practice there is considerable variation in the protective procedures that are used. There is also a wide choice of materials available, including an ever-extending range of non-metallic materials that are used for handling chemicals. These include glasses, hard rubbers and plastics. In building chemical plant, for example, the choice of material for containers and pipes is very wide. The economic factors are all-important, yet they are not always obvious. In some plants contamination of the product may be unimportant and a regular replacement of corroded sections may be cheaper than the initial use of a resistant alloy which will cause less contamination of the contents. In other plants, working continuously, delays in production resulting from the need to replace corroded parts may be so costly that the higher initial cost of a resistant material is relatively unimportant. Heavy chemical plant in prefabricated sections is expensive to transport and to assemble at any one place and these costs, which can arise every time a piece of equipment fails, will have an influence in determining the material used in construction. These economic points and many others have to be borne in mind in selecting the material for a large-scale chemical plant. Nor does any situation have constant conditions. Costs of metals and alloys vary from year to year. Political conditions are important both abroad, where the availability of an ore may suddenly change, and at home, where tax allowances on manufacturing equipment may be altered by a change of government. It may seem somewhat strange to mention such factors in a book on corrosion, yet any industrialist who is regularly engaged in building large-scale plant will only too readily confirm that many such completely non-technical factors do influence the choice of materials.

Corrosion problems during chemical manufacture and in engineering plant can be considerable. They cover most aspects of corrosion and nullifying their effect is sometimes referred to as corrosion engineering. It is concerned with the useful application of corrosion fundamentals to everyday problems, particularly large-scale plant.

Every type of corrosion problem can be found in chemical plant. A few can be indicated briefly. In chemical manufacture material is continually being made to flow and in the latter stages of such processes a concentration of the product is usually sought. Both the flow and the concentration produce rather special corrosion effects. In the latter case considerable potential differences may be established within a pipeline system where the concentration of the contents varies considerably. This can be appreciated by reference to the Nernst equation. It may be possible to overcome these potential differences by insulating the metal sections so that the large resistance of the corrosion circuit renders the corrosion current zero, by interposing an intermediate metal, or by using a more resistant material at the anodic end. Differential aeration effects, already described as a possible hazard producing pipeline attack, can also exert a damaging influence. If oxygen diffusion is rate controlling, then concentration differences may be of no great importance. On aluminium, for example, the oxygen may mend the film or depolarize the cathode, the effect being dependent upon the velocity of the medium[97]. This is another example of the dual role of oxygen described in section 3.4. Temperature gradients can also cause unexpected attack, for example, by depolarizing a useful controlling reaction. The state of dispersion of phases is important, particularly under critical condensation conditions where the establishment of thin films of moisture giving ready access of oxygen for cathodic reaction can cause marked increases in corrosion rate, as already noted in section 2.7 on atmospheric corrosion. Turbulence that may occur in a small section of a chemical plant can also cause considerable increases in attack.

The application of protective measures requires imaginative thinking. The throwing power of cathodic protection is fairly poor, particularly down long pipes. In one case cathodic protection has therefore been applied[97] by adding particles of magnesium of such a size that they are kept in suspension in the solution. Every time particles touch the oxide surface of the pipe it is locally protected and since this occurs many times the whole interior surface of the pipe is protected. On the other hand, protection of steel pipes in flowing water, using inhibitors, can be difficult, since any turbulence is likely to increase the corrosion rate, and the degree of inhibition, although determined as adequate under static conditions, may be quite insufficient.

The general problem in corrosion design demands a synthesis of cost, common sense, experience and corrosion theory.

Cost is an obvious factor. Many manufactured parts have an estimated useful life and the aim is to make them stand up to the intended environment for that time and not much longer. Therefore the aim must always be to use the cheapest materials. Cathodic protection is often provided for an imperfect coating of a pipeline since it is impracticable to specify or expect a flaw-free coating. By restricting the area to be protected, the power required is negligible compared to that required for a bare pipe. In painting, the labour cost is relatively so high that the best possible materials should be used. The cost of inhibiting a closed recirculating water system is obviously very different from the economic problems of a once-through water system. In the latter case, more expensive construction materials must be used or, perhaps, a cheap deoxygenating system, e.g. prior running of the water over scrap iron.

Common sense demands using knowledge of corrosion to the fullest extent while keeping a sense of perspective. There are obvious pitfalls. Avoid re-entrant corners and crevices, where lack of oxygen may produce a very active region. Where water can collect, make a drain hole. Watch out for dissimilar metal contact, and take care. In water containing dissolved oxygen, steel sheets joined by a copper rivet will survive, but copper sheets joined by a steel rivet will quickly fall apart since there is a very large efficient cathode available. Two stainless steels can develop noticeable galvanic current since there is a considerable range in the compositions available and corrosion potentials may differ quite markedly. It is possible for certain stainless steels to be passive and others active in the same environment.

Experience is the most difficult of all attributes to obtain. When faced with a problem in corrosion, there are some commonly occurring explanations which come flooding into one's mind. Galvanic attack, differential aeration, high chloride content, accumulation of active impurity on surface, some depolarizing effect, all these must be looked for. Experience is the art of sifting the possible explanations and discerning the correct one. Knowledge, practical ability, native shrewdness and intelligence are all necessary in solving corrosion problems, but experience is often the final arbiter.

Corrosion theory must be firmly understood. This is really the first necessity in attempting corrosion design. Experience and intuition are helpful attributes, but there is no adequate substitute for a basic understanding.

It is not only in chemical plant that good corrosion design is required. Every type of metal structure is likely to be subjected to some corrosive atmosphere. It is the marrying of the object to its corrosive environment with the aim of prolonging the useful life of the metal part or structure that makes corrosion design a considerable challenge. Much remains to be done in this field and it is likely that much progress will be made in the next few years.

CHAPTER 4

Corrosion Failures and Attack

4.1. INTRODUCTION

In this chapter some of the many types of corrosion attack that occur are described, starting with those that quickly result in the failure of metal parts. Other special forms are then described, including attack in the rather special environments of liquid metals and fused salts. After a brief description of the effect of some environments upon mechanical properties and an indication of the problems involved in corrosion testing programmes, the chapter and book are concluded with a final section that outlines the intricacies of common reactions that have been very simply described in earlier chapters and also covers some miscellaneous points.

Corrosion damage can have a serious effect upon the mechanical properties of a metal. When an oxide forms on a metal surface the amount of metal remaining is obviously reduced, and if this is under load, then the stress acting on it will be raised. The same is true if the metal is being dissolved. In either case if the ultimate tensile stress (U.T.S.) is exceeded, then the specimen will break. Apart from these simple considerations, however, there are types of metal failure that occur which either are found only under corrosive conditions or are considerably exacerbated by corrosive media. These constitute a major source of structural failures and may be listed:

(a) Stress corrosion cracking.
(b) Hydrogen embrittlement.

(c) Corrosion fatigue.

(d) Intergranular failure.

(e) Miscellaneous failures.

Before considering these phenomena, however, the general case of localized corrosion is further considered.

4.2. LOCALIZED CORROSION

Localized corrosion can be described as corrosion occurring at one part of a metal surface at a much higher rate than over the rest of the surface. It can take many forms[98]. It is of considerable importance, since many corrosion problems arise because although a suitable alloy may have been chosen for a particular environment in which it does not corrode generally it may be susceptible to localized corrosion.

Crevice corrosion has been described in section 2.1. It arises commonly from differential aeration, where some form of restricted oxygen access is responsible. Since the environmental changes produced can give rise to reactions that are different from those occurring anywhere else on the metal surface, the general phenomenon is referred to as occluded cell corrosion and will include crevice corrosion as well. Very small differences in potential can cause a considerable concentration of anions within a crevice[99]. Pitting corrosion, which is a further form of localized corrosion, is associated with the breakdown of a film and it occurs frequently on a completely flat surface. It occurs where the protective oxide film on a metal suffers only local attack and is otherwise stable in the solution to which the metal is exposed. The Pourbaix diagrams indicate the likely pH range, which is usually in the neutral/slightly alkaline range. Distinction can be made between pitting that arises from a differential aeration cell, similar to crevice corrosion, which many metals can suffer, and pitting that occurs in metals that exhibit passivity or are covered by oxides of high corrosion resistance, where any breaks that occur in completely unshielded conditions are likely to become sites of intensive attack.

Pits are nucleated wherever the oxide film is likely to be discontinuous and where the local environmental conditions are most suitable.

Emergent screw dislocations, along an emergent slip line, and sheared edges are likely sites, but it must be emphasized that the occurrence and distribution of corrosion pits is a complicated affair. In addition, pitting may be initiated by external factors, e.g. where external deposits have settled on a surface. Pits may be nucleated at points determined not only by faults in the surface film, but also at sites determined by the underlying metal. Such sites may arise from alloy heterogeneities in the surface, or be associated with grain or phase boundaries. Solid non-metallic inclusions arising from casting or processing procedures or from impurities, e.g. sulphides in stainless steels, may also provide initiation sites. The size of such metallurgical 'faults' may vary considerably from well under a micron, perhaps as small as a few atomic diameters, to millimetres, perhaps an inclusion visible to the unaided eye. Since the effect that such faults exert on the overlying oxide film is unlikely to be uniform, it cannot be expected that all 'faults' will cause pit initiation simultaneously.

A pit will often become covered with a membrane of corrosion product and if some of this falls down the face of a vertical specimen it may help initiate further pits, particularly if it becomes loosely attached to the surface and helps to create anaerobic conditions. Any material can do this, e.g. marine organisms or silt.

In steel pipes, which are often covered with millscale (magnetite), pits become covered with a membrane of ferric hydroxide. An equilibrium is established among the contents of the pits which have a pH of *ca.* 6.

On aluminium the reaction produces film-thickening, sometimes into the range of interference colours. The alkali formed during this reaction will tend to dissolve the film and not necessarily concentrate attack. If, however, the alkali is largely destroyed (e.g. by reaction with calcium bicarbonate) and if the local anodic products are not disturbed, then pitting may start, since the reaction produces hydrogen ions:

$$Al + 3\ H_2O = Al(OH)_3 + 3\ H^+ + 3\ \text{electrons.}$$

Unless these are destroyed or removed, the accumulation will cause a gradual lowering of pH. The attack, therefore, is gradually increased, so that the mechanism is autocatalytic. Eventually film formation will cease and the only reaction occurring at the base of the pit will be:

$$Al = Al^{3+} + 3 \text{ electrons.}$$

Waters containing calcium bicarbonate, oxygen, traces of copper (which may produce cathodes very much more efficient than alumina) and chloride ions (which interfere with film build-up) cause pitting on aluminium. The pH within a pit in aluminium has been measured as 3·5 in chloride solutions with a bulk pH of 7.

Pitting on copper is not likely, although particles and marine organisms may initiate it through differential aeration attack. If copper pipes are covered with a carbonaceous layer of low resistance, left on after drawing, then breaks in this may result in pits.

Stainless steels may pit extensively in the presence of chlorides and bromides in aerated solutions, where passivity is maintained. In de-aerated conditions pitting is not found, although heavy widespread attack may occur. Nitrates and alkalis generally inhibit the initiation of pitting although additions of nitrate will exacerbate already existing pits. Pitting is associated with a localized loss of passivity and the voltage drop across the active/passive region may be 0·5 V or more. A high current density will flow down the low resistance pit and provide cathodic protection of the adjacent area. Pits in metals and alloy that exhibit active/passive transition are much deeper than in non-passivating materials. Concentrated aqueous chlorides form in the pit and help maintain the heavy anodic dissolution rate at the bottom of the pit by preventing film formation.

Metal ions whose redox potentials fall within the passive range of a metal can exacerbate pitting by providing additional cathodic reduction species. Thus, while chlorides pit stainless steel in the presence of oxygen, only some, e.g. cupric and ferric chlorides, will also pit stainless steel in the absence of oxygen.

Pitting can occur on all metal surfaces. In practice it is found most commonly on passive alloys, where the very resistance of the film serves to maintain the attack in one place, rather than allowing it to spread. This will include metals like iron, which may have been rendered passive by an inhibitor which is not sufficiently high in concentration, a situation described in section 3.4. Pitting is prone to occur in certain environments, particularly chloride solutions. Pit initiation can take a long time to occur. In order to study the phenomenon it can be induced by anodic polarization. This procedure is also a useful method of comparing the pitting resistance of a number of alloys in a given environment. Specimens are polarized anodically and breakdown of the passive film is indicated by an increase in current at the breakdown potential, E_b. Such a result is shown in Fig. 74. The more noble the value of E_b, the better is the resistance of the alloy to pitting corrosion.

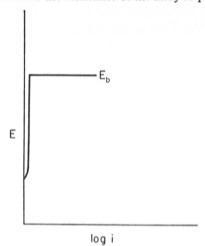

Fig. 74. Anodic polarization curve of a passive alloy in which breakdown of passivity occurs at E_b. If the full polarization curve is drawn from the active region to E_b the type of curve obtained is shown in Fig. 88.

If E_b is more active than E_{corr}, then pitting is likely to occur under open circuit conditions. E_b will tend to become more active as the pH is lowered and as the concentration of chloride ion is increased. E_{corr} will become more noble as the concentration of the cathodic species is increased, as explained in section 2.3, e.g. pH lowered or oxygen level raised, and such increases may induce pitting by making E_{corr} noble to E_b.

The determination of E_b is often done in a laboratory by a potentio-dynamic technique. A very slow scan rate is desirable, otherwise values of E_b may be obtained that are too noble. Some workers determine also a protection potential E_p by reversing the potential scan direction after breakdown so that the potential then changes in an active direction: E_p is the potential at which the fall of current occurs to its original value, corresponding to the re-formation of the passive film on the pit surfaces, a process referred to as repassivation. E_p is not always easy to determine, particularly if the pits are large and irregular in shape, since time is required to neutralize the acid within them and the length of this period will depend upon the volume of each pit.

When breakdown occurs above E_{corr}, at E_b under potentiostatic conditions, the current developed with the growing pits increases according to:

$$i = \text{constant} \times t^n$$

where t is the time from pit initiation and n is an integer.

There is some uncertainty about the value of n, which has been reported as being as low as 2[100] and as high as 5[101]. Several possible explanations have been put forward to account for these differences and it may be that each is true for separate examples. If all pits nucleate simultaneously, then it might be expected that $i = \text{constant} \times t^2$, since only the total area of each pit is growing. If pits are nucleated at different times, then n will be > 2 and possibly not an integer, since this will depend upon the range of sizes developed. It should be noted that if not all pits nucleate simultaneously, possibly because of a range of pit nucleating sites of different inherent tendency to break down, as indicated above, then E_b is likely to vary if the full range of faults is not found in each sample of a particular alloy. Minimizing non-metallic inclusions improves pitting resistance. A specimen not containing any inclusions in the surface would therefore exhibit a higher E_b than one in which inclusions were in the surface.

Some workers believe that E_b corresponds not to the initiation of pits but to that point at which the total current coming from all the growing pits becomes comparable in magnitude to the passivating current density and therefore noticeable. In that situation pitting would eventually occur at E_{corr} even though the measured E_b is noble to E_{corr}. While the exact significance of E_b is not entirely agreed upon, it does constitute a useful, practical measurement.

The shape of pits can vary considerably both under laboratory and under service conditions. In laboratory studies pits formed at low potentials grow slowly and may develop crystallographic faceting corresponding to the development of the most slowly dissolving faces. At very noble potentials round hemispherical pits with polished surfaces are sometimes observed. Under open circuit conditions and over a long period of growth, pits are often of a wide variety of irregular shapes, some causing undercutting where the pit diameter below a surface may be very much larger than the diameter observed on the surface.

4.3. STRESS CORROSION CRACKING

Many alloys when subjected to an external stress, or when left under residual stress, while in contact with certain corrosive environments, develop cracks which propagate in either an intergranular or transgranular mode. A list of susceptible alloys is given in Table XIII. The phenomenon is called stress corrosion cracking.

Confusion is sometimes caused by reference to 'stress corrosion embrittlement'. Specimens that crack do suffer a loss of ductility which therefore justifies the term 'embrittlement'. There are, however, differences of opinion about the nature of the propagating cracks, since, although the mechanism may in some alloys include bursts of brittle cracks, in other alloys this is unlikely. Consequently the term embrittlement refers only to the overall effect.

Stress corrosion cracking is a major form of corrosion failure. Much effort has been expended in attempts to understand and control it, but a complete solution has not yet been found. This is not altogether surprising, since it is a complex phenomenon which has mechanical, electrochemical and metallurgical factors. Before describing these, however, a few general points can be made.

TABLE XIII. SOME COMMON EXAMPLES OF STRESS CORROSION SYSTEMS

Alloy	Medium	Comments	Mode of failure (intergranular/ transgranular)
High strength steels	Water		I
High strength aluminium alloys	Chloride solutions, organic solvents	Appears to be due to moisture	I
Copper alloys— Cu–Zn, Cu–Al, Cu–Si α-phase	Ammoniacal solutions Some solutions amine		I/T
Magnesium alloys	Chloride solutions	May be due to moisture	I/T
Cu$_3$Au alloys	FeCl$_3$ solutions		I/T
Mild steels	Hydroxide and nitrate solutions	Phosphates and carbonates also cause cracking in some potential ranges	I/T
Austenitic stainless steels	Hot chloride solutions Hydroxide solutions		T/I
Zirconium alloys	FeCl$_3$ solutions Iodine at 350° C		I/T
Titanium alloys	Chloride solutions Organic solutions Fused chloride melts N$_2$O$_4$ Hot solid chlorides		I/T

The mode of failure in most stress corrosion systems can vary from being predominantly intergranular to being predominantly transgranular. Transition from one mode to the other may depend upon the stress level, heat treatment, the corrosive medium, the testing method, the temperature and other variables. The modes indicated in the table are intended to indicate the more common mode first and the less common one second, but the relative incidence of occurrence of the first to the second cannot be readily determined and is possibly different for each separate alloy in each medium. Not all alloys show transitions. A few aluminium alloys, for example, exhibit small amounts of transgranular cracking but since the majority do not, intergranular cracking has been shown as the only mode.

Firstly, stress corrosion cracking is generally found only in alloys, but care must be exercised in making such a distinction since inter-granular cracking of 99·999% Cu in an ammoniacal solution has been observed[102]. While this may be associated with the grain boundary impurities, i.e. the alloy content of the high purity metal, it is not usual to describe such a material as an alloy. Intergranular cracking of high purity iron has also been reported[103] and is caused by grain boundary impurities. Secondly, cracking occurs in alloys only in certain specific environments, e.g. α-brass in ammonia, as indicated in Table XIII, but the range of environments has been found increasingly to be greater than was thought originally. Where the cracking medium is water, the specific nature is not of much practical significance, since it can scarcely be avoided. Thirdly, stress corrosion cracking is a conjoint phenomenon. It occurs as a result of a stress being applied while a component is in a corrosive environment. Removal of either the stress or the environment will prevent crack initiation or cause the arrest of cracks that are already propagating. Fourthly, however the applied stress is derived, it must have a surface tensile component. Finally, the critical corrosion reaction responsible for crack propagation is not always clear. Cracking may arise from corrosion, i.e. metal loss occurring on a very narrow front, for reasons discussed below, but it may arise also from local embrittlement, arising from the absorption of hydrogen atoms that are discharged at local cathodes close to the crack tip. Sometimes a distinction is made between the two general mechanisms by referring to the first as an active path mechanism and to the second as a hydrogen embrittlement mechanism. While it is quite common to refer to them separately, in many examples of stress corrosion cracking the critical corrosion reaction is not known, so such distinctions cannot be made with confidence. It cannot be ruled out that *both* reactions may be necessary in some systems of stress corrosion cracking.

Before discussing features in individual alloy systems, it is of interest to consider separate general aspects of the problem arising from mechanical, electrochemical and metallurgical properties.

Mechanical Aspects

The total time to failure of a component in service or of a specimen in a laboratory test, t_f, is made up of a period of initiation, t_i, leading to the beginning of the crack which propagates during a period, t_p, at the end of which final fracture occurs.

Experimental techniques used for stress corrosion cracking studies include: (a) making U-bend specimens of the alloy and immersing them in the environment, a simple quick method which produces an unknown range of stress values in each specimen; (b) using tensile specimens, suitably enclosed in vessels with a bottom liquid seal, and measuring at least the time to failure, which can be done automatically. This method is readily adaptable to electrochemical investigations. Other common metallurgical techniques are also used, e.g. optical and electron metallography.

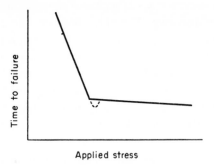

FIG. 75. Variation of time to failure and applied stress for austenitic stainless steels in 42 wt% $MgCl_2$ solutions[104]. The result indicated by the unbroken lines is found in other systems too. The deviation shown by the broken loop may apply to stainless steels.

The importance of the stress value in stress corrosion cracking is shown in Fig. 75, in which the time to failure for polycrystalline specimens is plotted against applied stress. This type of curve which exhibits a change of slope (a knee) is sometimes referred to, rather misleadingly, as a static fatigue curve. Such curves are found in stress corrosion studies of austenitic stainless steels[104]. For some of them the knee occurs at a stress corresponding to the approximate yield point, but for others the knee occurs at lower values of applied stress. The importance of yielding, however, is emphasized by experiments on single crystals of stainless steels which did not crack unless the applied stress was above the yield point[105]. There is some evidence that the point of intersection of the two lines in Fig. 75 has a trough or depression in it, as shown by the dotted line. It must also be noted that the t_f becomes very long at low stresses but no threshold stress is indicated below which cracking does not occur.

With high strength alloys, techniques of fracture mechanics have been employed. Specimens with notches and sharp fatigue cracks, like that[106] shown in Fig. 76, are stressed while in contact with the corrosive environment. In non-susceptible alloys fracture occurs only when

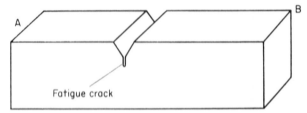

FIG. 76. Notched specimen pre-cracked by fatigue used for stress corrosion specimens. Specimens are clamped at one end and loaded at the other. Many forms of specimen can be used. Plane strain is obtained if the width is above a certain value which is dependent upon the material.

the plane strain fracture toughness is reached, K_{Ic}, where K is the stress intensity factor, which is derived from the crack length, c, and the stress, σ, acting across the narrow cross-section of the specimen so that:

$$K = k \times \sigma \times \sqrt{c}$$

where k is a constant dependent upon the configuration of the specimen. In susceptible alloys, stress corrosion crack propagation occurs (referred to as sub-critical cracking) until c has increased, so that K_{Ic} is exceeded and rapid overload failure then occurs. The initial stress intensity factor K is plotted against t_f as shown in Fig. 77. What is noticeable here is a value of K below which failure does not occur. This is designated K_{Iscc}. Where this exists it can be employed as a design criterion. To be certain that specimens will not break can be difficult

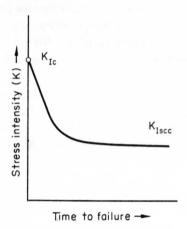

FIG. 77. Relationship between initial stress intensity factor, K, and time to failure t_f. In some systems a minimum value of K is observed below which failure does not occur. This is designated K_{Iscc}[113].

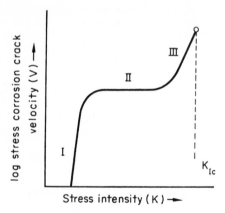

FIG. 78. General relationship between the stress intensity factor, K, and stress corrosion crack velocity V. Many systems exhibit only stages I and II[113].

since the velocity of cracking can be very low, e.g. 10^{-11} m/sec for aluminium alloys[107]. The velocity of stress corrosion cracks is dependent upon the value of K. The general relationship is drawn in Fig. 78, in which three regions of cracking are shown. Most alloys exhibit only stages I and II. Stage I is characterized by an activation energy of 112 kJ/mole (27 kcal/mole) in titanium and aluminium alloys and in high strength steels. Whatever processes are occurring in these alloys cease to be rate-controlling at the transition to stage II in which the velocity is constant and is controlled by diffusion within the environment of some species. The activation energy for this stage is 12–21 kJ/mole (3–5 kcal/mole) and for many alloys this represents a maximum velocity. It is dependent upon potential, viscosity, halide concentration and heat treatment. Stage III is observed only in extremely susceptible alloys and has been little investigated.

In lower strength, more ductile alloys fracture mechanics is less readily applicable. In cold-worked brasses the velocity has been shown[108] to be proportional to K^2, while in austenitic stainless steels in $MgCl_2$ solutions stress-dependent and stress-independent regions have been observed[109].

Electrochemical Aspects

Stress corrosion cracking can be regarded perhaps as the most extreme form of localized corrosion. If it is assumed that some of the highest velocities observed (cm/min) are caused by anodic dissolution, then the crack front will be dissolving at > 100 A/cm². Since mechanical tearing and cleavage occur also during fracture, such assumptions are not necessary and the contribution made by dissolution may be very small. What must be explained is why such dissolution occurs.

Stress corrosion cracking occurs most commonly in alloys that are employed only because they are normally covered with a highly protective oxide layer, e.g. aluminium and titanium alloys and stainless steels. Once breakdown has occurred, it can be argued that continual plastic deformation at the tip of a stress corrosion crack prevents the complete repassivation of the metal surface in the region of the crack tip. The potential difference between a filmed side and an unfilmed tip could be quite large[110] (0·5 V). The important requirement is considered to be that the repassivation process should occur within a fairly narrow time interval. Such a thesis separates the general process of film rupture, that can occur on any passive alloy, from the particular process that produces susceptibility. The significance of the repair process is emphasized by the schematic drawing[111] of Fig. 79. An emergent slip step breaks the thin protective film and exposes a clean fresh reactive metal surface to the solution. Since the metal is passive, it can be expected that the

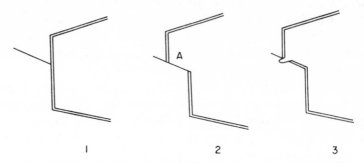

FIG. 79. Schematic diagram depicting in three stages the creation of a slip step on an alloy surface covered with a passive film. The freshly created surface will be rapidly refilmed but during the time in which this occurs corrosion attack and hydrogen entry are both likely to occur. Under very specific conditions the delay in repassivation will result in crack nucleation. Under a tensile stress the slip plane in (1) breaks the protective film, creating unfilmed metal surface (2) which repassivates except for a small part which undergoes dissolution and possibly exhibits hydrogen absorption (3)[111].

freshly generated surface will repassivate, but during the time that this takes to occur two processes will occur. Some dissolution will occur and some hydrogen absorption will occur. A process of competitive adsorption between species promoting passivity, e.g. OH^-, and species promoting activity, e.g. Cl^-, is envisaged. If repassivation occurs too rapidly, insufficient corrosion attack occurs to cause an increment of crack growth, and if repassivation occurs too slowly, then too much corrosion attack occurs and some form of elongated pit forms, rather than a narrow crack increment. Thus it is hypothesized that there will be a narrow range of repassivation times which will result in crack propagation.

It is important to regard the situation depicted in Fig. 79 as only schematic. In practice, deformation occurring at the crack tip will not take the form of a single step. But Fig. 79 does provide some insight into the kinetic processes of deformation and repassivation. A general hypothesis can be made: propagation occurs as a result of more fresh metal surface being generated than can be repassivated adequately within a given time.

An example[112] of the importance of repassivation is shown in the results illustrated in Fig. 80. Specimens of a susceptible Ti–5 Al–2·5 Sn

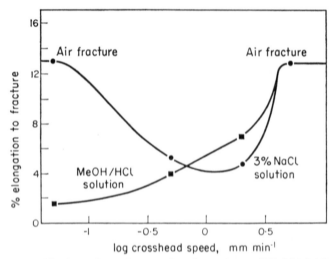

FIG. 80. The elongation to fracture of tensile specimens of Ti–5 Al–2·5 Sn alloy of 3 cm gauge length in 3% aqueous NaCl and in MeOH +1% HCl environments as a function of Instron crosshead speed[112].

alloy were strained at several strain-rates while in contact with two different solutions: aqueous 3% NaCl in which titanium alloys are passive, and a methanol/1% HCl mixture in which titanium alloys corrode. At high crosshead speeds (corresponding to a high strain-rate), ductile fracture occurs since there is insufficient time for crack initiation. At lower crosshead speeds, crack initiation and propagation occur in both solutions and the total elongation to fracture is diminished as a consequence. At the lowest crosshead speeds that were employed, two different results were obtained. In the corrosive mixture the elongation diminished since there was longer time for propagation before the ductile overload failure point was reached. In the aqueous solution, however, no cracking was observed because the amount of fresh surface being generated was too low and repassivation was possible. The ductility observed was therefore the same as at the high crosshead speeds or in air.

Polarization has usually a marked effect upon t_f. Anodic polarization will usually shorten t_i very much, lower t_p by increasing stage II, and it may increase the crack density by up to 10^2. Cathodic polarization will usually lengthen t_f and may prevent failure altogether by cathodic protection. If hydrogen is the embrittling species, anodic polarization may still shorten t_f if hydrogen discharge inside the pits or crack nuclei increases, a point that is discussed in section 4.9. Cathodic polarization can be expected to shorten t_f, since hydrogen is

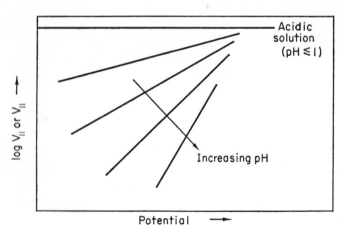

FIG. 81. Schematic drawing of the relationship between stage II velocity in titanium alloys as a function of pH[113].

more readily available, but this will not be true if the surface absorption of hydrogen is rate controlling when no change in t_f may occur. An increase in t_f can be expected if the alkalinity generated at the cathode promotes the formation of a protective film which is a high resistance barrier to hydrogen ingress, as, for example, in titanium alloys. This point is illustrated in Fig. 81 in which the effect of potential and pH on the stage II velocity is shown[113] for titanium alloys in aqueous chloride solutions. Where film formation is possible, cathodic protection is possible, but where film formation is not possible (pH 0), cathodic protection is not obtained.

The exact process of initiation has not been examined closely. In not very aggressive environments initiation may be associated with visible pits whose function is mainly chemical, i.e. to produce the necessary acidity by hydrolysis for crack propagation, or mainly physical, i.e. to act as a stress raiser. The acidity developed inside cracks has been examined by freezing the solution and then analyzing it and by electrochemical techniques[114]. In titanium alloys in neutral aqueous NaCl the pH at the tip may be as low as 1·7, while in aluminium alloys it is 3·5. In high strength steels the value has been shown to be 4 under conditions where the external pH of the bulk solution is between 2 and 11, a result that emphasizes the importance of the hydrolysis reaction. Occluded cell reactions give rise to acidity, but the level of acidity is determined from the hydrolysis constant of the appropriate metal halide salt. In some alloys the pH produced by hydrolysis at the tip of the crack is alkaline, e.g. magnesium alloys.

Metallurgical Aspects

Many metallurgical variables have an effect upon stress corrosion susceptibility. Susceptibility is usually reduced as the grain size is diminished. Mechanical strength is of major importance. Cold work usually increases susceptibility. Heat treatment affects susceptibility, which is often greatest at the peak strength of an alloy and diminishes as that strength is lowered. An example[107] of the effect of heat treatment on crack velocity in aluminium alloys is shown in Fig. 82.

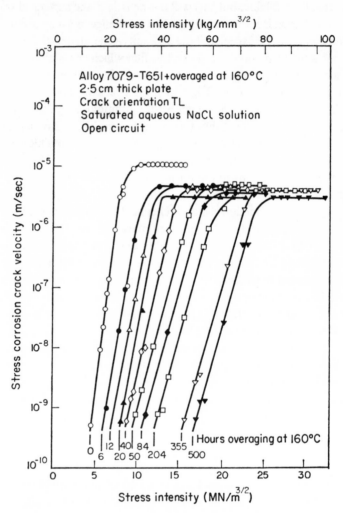

FIG. 82. The effect of overaging on the stress corrosion crack velocity of a high strength aluminium alloy[107]. The relative changes in stages I and II cracking vary from alloy to alloy.

The deformation modes of alloys, as revealed by transmission electron microscopy, are relevant to stress corrosion susceptibility. Face-centred cubic alloys of austenite and α-brass, both of which exhibit transgranular alloys, develop co-planar arrays of dislocations when lightly deformed. Other fcc alloys that exhibit similar patterns do not crack, so that the mode is not a sufficient cause of susceptibility. Hexagonal α-titanium alloys containing aluminium or oxygen exhibit similar co-planar arrays and undergo transgranular cleavage. Co-planar arrays represent a restricted slip mode, which can be the result of a low stacking fault energy, short-range order, or other unspecified causes. Aluminium alloys that are susceptible to intergranular cracking when aged near to peak hardness and maximum susceptibility exhibit restricted slip of a different kind. Dislocation bands are formed which contain a high density of dislocations. Lowered susceptibility from overaging is accompanied by a more dispersed arrangement of dislocations.

Mechanisms of Stress Corrosion Cracking

While many sequences have been proposed to account for the propagation of stress corrosion cracks, it is possible to reduce them to five major divisions.

(i) *Dissolution of yielding material*

This has already been described and arises from inadequate repassivation. In addition, it has been suggested that microstructural features, such as stacking faults or the regions immediately adjacent to them, are preferentially corroded, probably as a result of solute segregation on an atomistic scale.

(ii) *Fracture of a corrosion product film*

The surface is converted to a corrosion product film, possibly to a greater extent over certain features, e.g. grain boundaries. Under the influence of the external stress it fractures at the grain boundaries and re-forms there. This then becomes a repeating sequence accounting for propagation and giving rise to an intergranular fracture. It may be possible for transgranular cracking to occur in this way by emergent slip steps causing the repeated fracture of the film. In some situations corrosion product within the crack has been described as causing crack extension by a wedging action arising from metal being converted to a more voluminous corrosion product.

(iii) *Hydrogen embrittlement*

This description covers several different possible mechanisms. Hydrogen is absorbed at the tip of the crack at a rate faster than it can disperse from that region by diffusing into the bulk of the specimen. In one of several possible ways it causes embrittlement of a small volume of metal which cracks under the influence of the external stress.

(iv) *Stress sorption*

Fracture occurs as a result of a species being adsorbed from the solution and causing bond weakening at the crack tip as depicted in Fig. 83. The species could be hydrogen atoms and this would then constitute another form of hydrogen embrittlement.

(v) *Intergranular cracking*

This may occur as a result of selective corrosive attack on grain boundaries. The function of the stress may be merely to aid separation and it may not influence the rate of the chemical reaction. The narrow path of selective corrosion pre-exists before the stress is applied. This is probably the simplest mechanism of cracking.

Early theories of stress corrosion cracking were concerned with a two-stage process: firstly, the electrochemical reaction produces a stress-raising pit and from this, secondly, a crack runs a short distance

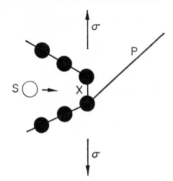

FIG. 83. Schematic diagram of an atomically sharp crack under a tensile stress σ. Cleavage fracture occurs when the atom–atom bond X is broken. Inside the material shear may occur on the slip plane P. The strength of X may be lowered by the adsorption of a species S from the environment. Absorption may raise the shear stress on P. If the ratio of cleavage stress/shear stress is lowered, then there may be a change in mode of failure from ductile to cleavage.

and the electrochemical reaction then occurs again. For mild steel in nitrate solutions and for some aluminium alloys this two-stage propagation has been detected by measuring sudden 'kicks' in the potential of specimens, irregular extension of specimens (difficult to interpret in specimens containing large numbers of cracks), and by acoustical methods. In austenitic stainless steels the two-stage process has not been detected. Irregular crack propagation can be explained in terms of grain boundary precipitates in mild steel or in association with certain intermetallic compounds in some aluminium alloys, whereas austenitic stainless steels are highly ductile alloys and unlikely to be able to sustain a stress-raising notch which could produce a short brittle crack, since plastic relaxation would seem likely to blunt the tip. A similar objection could be made for the stress corrosion cracking of α-brasses, although in these it has been argued that regions of local short-range order may sustain brittle cracks[115].

In 18 Cr–8 Ni austenitic steels, which crack in the presence of chloride ions, it has been suggested that the mechanism of cracking is a one-stage process·[104]. The rapidly yielding material at the tip of the crack is continually dissolved by the chloride ion. Cracks propagate through this material at rates as fast as 1 mm/hr and the anodic current density required to dissolve material equivalent to this rate is *ca.* 1 A/cm². Electrochemical measurements have shown that the potential at the tip of a crack in 18 Cr–8 Ni steels in 42 wt% $MgCl_2$ solution boiling at 154 °C (which is a common testing medium since it quickly induces cracking) is − 150 mV E_H approximately, whereas an imposed current density of 1 A/cm² under ordinary static conditions raises the potential by at least 750 mV above that value. It has been hypothesized that at the tip of a crack there is not only a very high strain rate but also a continual renewal of fresh solution. These conditions were simulated over a length of wire, anodically polarized at 0·5 A/cm², by straining it and simultaneously flowing the corrodent past[116]. Under these conditions the large polarization potential was removed. The work was also done potentiostatically at − 150 mV and the current density increased by more than × 10⁴[117]. The significance of these results is that this very considerable increase in dissolution rate when the steel was being strained was only found for 18 Cr–8 Ni alloys in the concentrated chloride solution. Materials that are not susceptible to cracking in chloride solutions, e.g. iron, exhibited very little sensitivity and 18 Cr–8 Ni steel exhibited little in sulphate solutions, in which it does not crack.

Analogous results were obtained for a series of Fe–Ni alloys of very high purity. Fe–5% Ni and Fe–10% Ni alloys, which both cracked in the 42 wt% $MgCl_2$ solution, showed a considerable increase in dissolution rate when subjected to a strain rate under potentiostatic investigations. Conditions similar to those employed for 18 Cr–8 Ni steels showed increases > 10³ times over the static value while non-cracking alloys, e.g. 50 Fe–50 Ni, showed an increase of only × 10[118].

The explanation for the very rapid attack under yielding conditions in susceptible stainless steels is thought to be connected with certain microstructural features in the face-centred cubic austenite lattice. It is a characteristic of these alloys that they have very low stacking fault energies and very large numbers of dislocations on slip planes. Electron microscopic studies have shown[119] that the specific environment does attack these large pile-ups almost exclusively and it is possible that this might explain the connection between strain rate and dissolution rate. The reason for this is not clear, but there is some evidence that microsegregation occurs in the region of large pile-ups and that this makes either the pile-up or the region adjacent to it particularly active to corrosion. Both susceptible austenitic stainless steels and α-brasses are alloys that have low stacking fault energies and suffer transgranular cracking. Other copper alloys fail intergranularly in ammoniacal solutions, e.g. dilute Cu–P, Cu–Si, Cu–Al alloys, and while little fundamental research has been done on them, the failure of cracks to penetrate into the grains appears to be associated with the high stacking fault energies of these alloys[119].

Impressed cathodic polarization has also been used both to delay and eliminate the onset of cracking and this does represent a universal form of protection from stress corrosion cracking for all systems. Deoxygenation of the solution is also a means of reducing stress corrosion cracking where oxygen reduction is the major cathodic reaction. Inhibition has been employed to protect stainless steels from cracking. Breakdown products of urotropine[119] and large concentrations of sodium phosphate[120] have both been successfully employed.

The highly concentrated chloride solution used in many laboratory investigations is not met with in practice. Service failures occur in austenitic steels in pressurized water containing as little as a few ppm of chloride ion and a few ppm of dissolved oxygen at about 200–300 °C and above. Often the cracks are associated with a white corrosion product, which is not found in the laboratory environment. It may be that in very dilute environments pit development and associated visible corrosion product are necessary in order to achieve the necessary acidity. In service failures cracks are sometimes found filled with corrosion product and it has been suggested that these serve to 'wedge open' cracks[121].

Hydrogen is evolved during the cracking of austenitic steels, although most of the cathodic reaction is the reduction of oxygen. Since cathodic polarization stops cracking, it appears that anodic dissolution is a factor in the crack propagation mechanism.

For a given alloy, variation in small alloying additions may have a pronounced effect upon susceptibility to stress corrosion cracking. Since the average commercial austenitic steel contains many elements, the determination of the effect of variation of any one is not easy. Some results have been usefully subjected to a regression analysis using a computer[122]. The interaction of elements must also be taken into account. For example, purified 16 Ni–20 Cr alloy is not susceptible to cracking; the addition of 1·5 Mo renders it susceptible[123].

The significance of these effects is probably related to the microstructural changes that they cause and, in particular, to the stability of the austenitic lattice.

For the brasses, lowering the zinc content increases the time to failure under stress corrosion cracking conditions and it may be noted that this also raises the stacking fault energy of the alloy. Such effects in other alloys have not been investigated so fully. It has been suggested that cracks in α-brass may progress by the formation of a tarnish film which fractures and then re-forms, etc.[124].

Mild steel undergoes cracking under alkaline high temperature conditions, referred to as caustic cracking (to which austenitic steels are also susceptible but to a much lesser extent than mild steel). It represents a serious hazard in boilers and steam equipment where crevices or porous scale in highly stressed regions, e.g. a rivet or seam, can become regions of high pH and initiate explosive failure. Cracks are intergranular, although there is attack on pearlitic cementite, and the iron dissolves as oxyanions. Magnetite is precipitated, but usually away from the crack tip, so that there is no stifling. Hydrogen is produced during the attack and it may help crack propagation by forming internal blisters.

Stress relieving of mild steel structures helps to reduce the probability of cracking, but this is usually combined with inhibitive methods aimed at eliminating the possibility of alkaline concentration build-up. Phosphate, which precipitates at high pH but below dangerous values, is widely used and added so that the ratio $Na_2O : P_2O_5$ is just less than 3 : 1. Sodium nitrate is another additive that has been successfully employed to eliminate cracking on locomotive boilers. Under some circumstances nitrate may accelerate cracking.

Age-hardenable aluminium alloys are subject to stress corrosion cracking in aqueous environments or in organic environments containing traces of water. The cracking is almost entirely intergranular and is accelerated by Cl^-, Br^- and I^- ions but not by F^- ions. Maximum susceptibility occurs near to peak hardness. The effect of overaging is shown in Fig. 82. In addition to deformation modes, as already discussed, grain shape is important. In rolled sheet, for example, maximum susceptibility is usually observed in the short transverse direction. The shape, structure and density of the intergranular precipitates and the zones adjacent to them are also of importance. While there has been a tendency to assume that the cracking mechanism is of the active path type, with pitting as a precursor and the electrochemical characteristics of the precipitates being of importance, recent work[125] has suggested that absorbed hydrogen may be of importance Thus the relative humidity controls the velocity, as shown in Fig. 84. At the very low values of relative humidity there is no pool of water at the crack tip. Each water molecule reacts with the metal as it arrives at the crack tip. It has been demonstrated[125] that hydrogen enters an aluminium alloy in sufficient amount to embrittle it under dynamic stressing by a mechanism characteristic of a strain-aging phenomenon. Temperature and strain rate dependence and the reversibility of ductility with respect to the hydrogen content all indicate that hydrogen plays an important role. The surface reactions appear to be rate controlling in the hydrogen absorption process. The beneficial effects of the overaging treatments may arise from creating a microstructure which has a greater tolerance for hydrogen.

Titanium alloys can fail by stress corrosion cracking in laboratory tests in a very wide variety of environments, as is indicated in Table XIII. Under service conditions, however, very few failures have been reported, a feature which can he explained partly by the difficulty of initiating cracking since titanium is usually covered with a highly protective film. At room temperature, hexagonal α-alloys based upon Ti–Al and Ti–O alloys fail by transgranular cleavage in aqueous NaCl solutions. Crack initiation is difficult because of the rapid repassivation and is achieved by dynamic straining tests as described in Fig. 80, or by

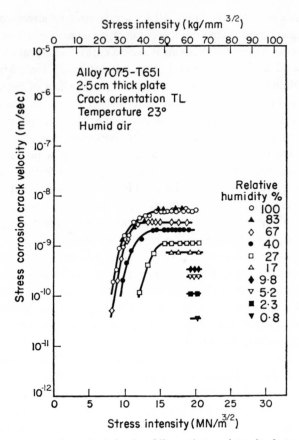

FIG. 84. The effect of relative humidity and stress intensity factor on the stress corrosion crack velocity in a high strength aluminium alloy[107].

the use of pre-cracked specimens. Similar fracture occurs in a wide range of organic solvents and it is not clear whether residual water is responsible for it. The bcc β-phase containing V or Mo is not susceptible and important commercial alloys like Ti–8 Al–1 Mo–1 V and Ti–6 Al–4 V consist of arrangements of α- and β-grains that depend upon heat treatment. β-phase alloys based upon Mn cleave along [100] planes.

While an active path mechanism has been advocated, the maximum velocities observed ($>$ 1 cm/min) require very high dissolution rates ($>$ 100 A/cm^2). There is some evidence that these can be achieved. Alternatively, it has been argued that absorbed hydrogen is responsible for the cleavage which is a mainly mechanical fracture as can be seen from fractographic examination. It has been shown[126] that cathodic charging followed by fracture can produce fractures similar to those obtained during stress corrosion, but that these are not observed if specimens are aged between charging and fracture.

Cracking is also observed in N_2O_4, possibly by the successive rupture of a corrosion product film. It is avoided by ensuring that NO is present in the N_2O_4. Cracking in red fuming HNO_3 has also been known for many years. In MeOH/HCl and MeOH/I_2 mixtures, intergranular stress corrosion occurs, probably by an active path mechanism. Grain boundaries in all α-alloys are selectively corroded, even in the absence of a stress. In MeOH/HCl there is a transition from intergranular to transgranular, at some value of stress, corresponding to a change from the relatively slow intergranular dissolution to the relatively rapid transgranular mechanical stress corrosion.

At higher temperatures hot salt cracking is observed in all stressed titanium alloys. This occurs $>$ 250 °C and requires a chloride, oxygen and moisture. Cracking is both transgranular and intergranular. Several mechanisms have been proposed based upon the localized production of Cl_2 or HCl. Considerable hydrogen pick-up occurs and this may be an important factor in establishing cracking. Thus hydrogen contents in fracture surfaces in excess of 10,000 ppm have been found in alloys with a residual hydrogen level of 70 ppm. These amounts were observed after tests in atmospheres with a moisture level of 0·25 ppm and were not affected by increasing this level by 10^4 times, a result that emphasizes the rate controlling action of the surface.

Molten salt stress corrosion has been reported with phenomenological features very similar to room temperature aqueous cracking.

4.4. HYDROGEN CRACKING

Some alloys fail under stress in corrosive conditions because of the entry of hydrogen atoms into the alloy lattice. This phenomenon is called hydrogen cracking or hydrogen embrittlement.

Stress corrosion cracking, which relies upon anodic dissolution for at least a part of the propagation process, can be avoided by cathodic protection. Hydrogen cracking, by contrast, is induced by cathodic polarization, since this promotes the production of hydrogen. This distinction has been used to discriminate between the two mechanisms.

Hydrogen cracking occurs commonly in ferritic and martensitic steels in sulphide environments, e.g. in association with oil well operations. The sulphide ion interferes with the evolution of hydrogen which migrates, probably in atomic form, into the metal.

The effect of hydrogen has been investigated by cathodic charging experiments.

The hydrogen appears to have no effect upon the elastic modulus of the material. Embrittlement is a reversible process, since a specimen charged with hydrogen will gradually lose it if left. At very low temperatures embrittlement by hydrogen does not occur. The amount of hydrogen that a lattice can hold may be well above the solubility limit ($>$ 100 times), particularly if the material is cold worked. The excess hydrogen collects in interfaces between solid non-metallic inclusions and the metal matrix and in voids. Under the action of the externally applied stress, the hydrogen concentrates in regions associated with localized regions of triaxial strain within the lattice.

Cracks are mainly intergranular and start where the surface is subjected to the highest tensile stress.

Severely cold worked austenitic steels can also suffer hydrogen cracking in addition to stress corrosion cracking. The application of either anodic or cathodic polarization shortens the time to failure of such material. Very high strength steels are particularly prone to embrittlement and extreme care must be exercised in welding them, etc. Some even fail if stressed in ordinary humid atmospheres. Protective methods, e.g. plating, can be harmful, since they may promote the entry of hydrogen. Special care must therefore be exercised at all times, e.g. phosphating should be accelerated. One outstanding feature of hydrogen embrittlement is that it is only detectable at low strain rates and standard notched-bar tests will not indicate the effect. Since it is likely that the hydrogen in the lattice diffuses to the deformed region at the tip of the crack, a high strain rate does not allow the necessary time for this movement to occur.

It is not always easy to distinguish between stress corrosion cracking and hydrogen embrittlement, particularly in ferritic and martensitic materials. Cathodic polarization, as already explained, should make a clear distinction, but results are not always conclusive. Fractographic analyses and the potential dependence of crack velocity both indicate that high strength steels fail by hydrogen embrittlement even under anodic polarization. Martensitic steels, in high temperature water, for example, may develop pits or stress-raising notches electrochemically, but the actual cracks, which run straight and unbranching across specimens, perpendicular to the applied stress, may propagate by some form of hydrogen embrittlement. Such cracks grow very rapidly. Specimens can actually be heard cracking.

Such rapid failures in ferrous materials, arising from hydrogen embrittlement or perhaps caustic cracking, constitute a serious hazard. Catastrophic failure can occur with no prior warning.With non-ferrous materials cracking occurs in specific media, which can either be avoided or used with safe alloys, although these measures can be costly. With ferrous materials, particularly high strength steels, failure can occur from embrittlement by hydrogen, which is derived from any aqueous environment. It is thus much harder to avoid.

While it is possible that high pressure molecular hydrogen exists in voids, current opinion and evidence suggest that atomic hydrogen migrates to the straining region where it exerts an embrittling effect. The mechanism is not clear. Possibly the hydrogen lowers the bonding of the metal lattice at the tip of the crack. It may collect on dislocation sites and prevent plastic deformation so that the fracture stress at the tip of the crack is reached before any yielding can occur.

4.5. CORROSION FATIGUE

When a metal is subjected to an alternating or fluctuating stress it is liable to develop cracks that gradually propagate through the material. The fracture is called a fatigue failure.

The presence of a corrosive environment hastens the cracking process. In Fig. 85 the mean applied stress vs. number of cycles to failure, commonly referred to as the S/N curve, is shown for a metal in a non-corrosive environment and then for the same metal in an aggressive

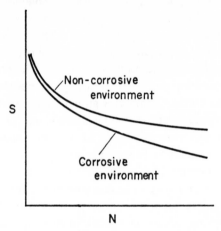

FIG. 85. Typical shape of an S/N curve for a metal in a non-corrosive and a corrosive environment under identical testing conditions.

environment, in which it fails more readily. The fatigue life of copper, for example, is much longer in a vacuum than in oxygen, while that of aluminium is shortened by the presence of moisture. The effect can be detected even in environments that are only very mildly aggressive.

During the fatigue process the metal slides along glide planes where cracks are nucleated after only *ca.* 5% of the total life. It is possible that emerging slip steps expose fresh bare metal that is rapidly oxidized and when the cycle is reversed this oxide layer is pulled into the metal where it interferes with the deformation process, shortening the time

to failure and aiding the normal metallurgical fatigue process. Therefore fatigue in ordinary air is a mild form of corrosion fatigue, but it
must be emphasized that fatigue does occur in completely non-
aggressive atmospheres.

In aqueous solutions corrosion fatigue is increased by anodic
polarization and retarded by cathodic polarization, unless the material
is susceptible to hydrogen embrittlement. The fresh emerging slip
steps represent very active anodic sites rather similar to the situation at
the tip of a stress corrosion crack. Anodic inhibitors have been used to
eliminate fatigue, but in practice this can be dangerous if there are
recessed regions that are not reached by the inhibitor. Cracks may
develop there and the stress concentration at the tips of the small
number of cracks that is formed may be greater than if the whole
surface were covered with cracks. Under these conditions of incom-
plete inhibition the fatigue life of specimens may be shortened.

In practice, corrosion fatigue occurs under many circumstances.
Superheaters, for example, are prone to fail where a barrier of steam
between the boiler wall and water in the tube is non-wetting. The wall
temperature rises until the steam film collapses and brings the cooler
water and pipe into contact. The fluctuating wall temperature creates
fatigue conditions. Caustic cracking and hydrogen cracking may occur
under these same conditions.

The fatigue life of metals is most commonly prolonged by putting
the surface under a compressive stress, e.g. shot peening, and corrosion
fatigue is reduced by plating, e.g. zinc on steel, cladding, e.g. aluminium
alloys, painting and by intelligent design, e.g. avoiding crevices where
pits may be initiated, and stress-raising notches. If pitting occurs on
surfaces that have been put into a state of compressive stress, then the
advantage of this treatment is likely to be lost when the depth of pitting
becomes comparable to the depth of the compressed layer.

Protection against fatigue must be used with caution. Electroplated
layers are frequently under tensile stresses, whereas it is compressional
surface stresses that improve fatigue life. Special plating baths have
been developed that deposit layers that are under compressional
stresses. Some layers may give galvanic protection, yet if the cathodic

reaction is hydrogen evolution, as in acid environments or with very active metals, this may enter the metal and shorten the fatigue life. For the same reason cathodic protection is not very effective in acid environments and may even be harmful. Anodizing appears to shorten the fatigue life of aluminium. Electropolishing has also been found disadvantageous. In applying metal or chemical coatings or paints, discontinuities must be guarded against lest heavy attack be established on a small number of sites. This is always a consideration in all forms of protection, but fatigue presents special problems since the surface is continually being deformed. The physical properties of coatings are of importance for the same reason.

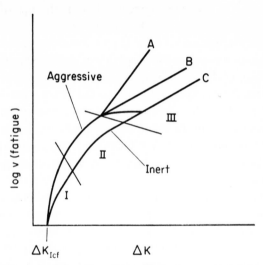

FIG. 86. General relationships observed between corrosion fatigue crack velocity and stress intensity factor cycle ΔK[127].

Recent investigations[127] of the variables controlling crack velocity have described a situation that is somewhat similar to stress corrosion cracking. For specimens tested under sinusoidal loading conditions with a zero average stress, three regions of cracking are observed that are dependent upon the maximum deflection that is employed which determines the cyclic stress intensity range ΔK. The general relationship between crack velocity and ΔK is shown in Fig. 86 for two environments, which are inert and aggressive. A general formula can be derived from the results:

$$\frac{da}{dN} = \text{constant} \ (\Delta K)^n$$

where da/dN is the crack increment for each fatigue cycle and n is an integer. At low values of ΔK, n may reach large values, e.g. 150, and the curve becomes vertical, in which case a ΔK_{Icf} is observed, an environment sensitive feature analogous to K_{Iscc}. Below the value of ΔK_{Icf} fatigue is not enhanced by the presence of a corrosive environment. Stage II has a strong environmental dependence and it is less stress dependent than stage I. In some systems of corrosion fatigue, e.g. titanium in NaCl, a plateau may exist, but the fatigue process in the absence of an environmental effect, e.g. in dry argon or in a vacuum, is always stress dependent. In many examples n has a value of 4. In stage III, three different modes of behaviour may be observed as indicated in the general form of Fig. 86. Firstly, the presence of an environment produces an additional accelerating effect corresponding to curve A. Secondly, there may be no additional effect as with curve B which remains above the curve for the inert environment C but parallel to it. Thirdly, there may be a convergence of the corrosion fatigue B to C, indicating the disappearance of the corrosion fatigue effect at high values of ΔK.

These are very general points for what is a very complex phenomenon and there is considerable variation from one alloy to another. In addition, the possibility of getting stress corrosion as part of the corrosion fatigue process in alloys susceptible to stress corrosion must be considered. In stage III there is some evidence[127] that corrosion fatigue and stress corrosion cracking can be additive in effect.

4.6. LIQUID METAL EMBRITTLEMENT

Many metals and alloys exhibit embrittlement when exposed to other liquid metals. The general requirements[128] for this phenomenon to occur appear to be (i) a tensile stress, (ii) a pre-existing crack or some amount of plastic deformation, so that dislocation pile-ups against an obstacle can occur, resulting in a concentrated tensile stress across the potential fracture plane, and (iii) adsorption of the embrittling species at the obstacle. Limited mutual solubility is required too, since too great a solubility may result in dissolution, with the possibilities of crack blunting. Little tendency to form intermetallic compounds is also a requirement, since this is a characteristic of high bonding strength. Velocities as high as 500 cm/sec have been reported. The proposed mechanism requires the adsorption of the metal at the strained bond at the crack tip. The redistribution of electronic bonding resulting from the process of adsorption results in bond weakening and fracture. Due to the nature of the metallic state, any such electronic effect is purely a surface one, since the interior layers are screened by the mobile electrons of the atoms.

In agreement with such an analysis, an element known to form high melting point intermetallic compounds can cause the inhibition of liquid metal embrittlement, e.g. 0.4% a/o Ba added to mercury prevents the embrittlement of aluminium[129].

Liquid metal embrittlement occurs more readily in an alloy as the stacking fault energy is lowered[130]. Since τ increases as the stacking fault energy is lowered, this would lower the ratio σ_c/τ, in accordance with the requirements described in Fig. 83. In addition, it has been suggested that such alloying might be expected to lower σ_c, as a result of altering the electron/atom ratio.

In fcc metals the transition temperature from ductile to cleavage in liquid metal environments depends upon the environment and not upon the material. The effect, for example, of temperature upon the yield stress or the metallurgical transition is not important. This is confirmed in Fig. 87 in which the strain at fracture is shown as a function of temperature for aluminium in a number of mercury/gallium solutions[131]. Recently practical examples have been reported in which the embrittling metal was in the solid state, viz. cadmium-plated titanium fasteners, which exhibit intergranular fracture. This is not unlikely provided that the two metals are in direct physical contact.

Some alloys, e.g. aluminium alloys, exhibit stage I and II cracking

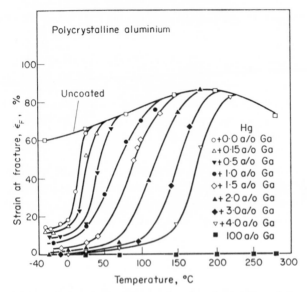

Fig. 87. The temperature dependence of strain at fracture for polycrystalline aluminium specimens in mercury, gallium and mercury–gallium solutions[131]. The results indicate that the ductile/brittle transition temperature is determined by the environment and not by the metal.

corresponding to Fig. 78. By contrast with stress corrosion cracking in aqueous media, the slope of stage I is much steeper with liquid metal embrittlement than with aqueous solutions and the maximum velocities much higher (as high as 500 cm/sec). Clearly the mercury must reach the tip of the crack, and where this is the slowest process cracking can be increased by raising the hydrostatic pressure on the mercury, thereby forcing it into the crack[132].

Monocrystals also suffer from mercury cracking, e.g. zinc. It is likely that mercury penetrates down subgrain boundaries or dislocation pipes. In practice, dilute solutions of mercurous nitrate are used to test how well cold-drawn brass tubes have been stress-relieved. If this has not been adequately done, the residual cold work will produce cracking.

4.7. OTHER FORMS OF CORROSION ATTACK

Environmental Effects on Mechanical Properties

In addition to corrosion cracking and corrosion fatigue, there are other effects that environments have on the properties of metals. Since these are largely concerned with alterations to the work hardening characteristics most experimental observations have been made on single crystals.

Firstly the presence of an oxide film can raise the yield point of a material, probably by preventing the emergence of surface dislocations as slip steps. The creep rate of heavily oxidized single crystals of zinc, for example, is increased considerably if they are immersed in dilute hydrochloric acid[133]. A specimen that had been etched in hydrochloric acid did not show any change in creep rate when acid was introduced.

Polar molecules combine with metals to form soaps, as already noted in section 3.6. The effect that they have upon mechanical properties has been disputed. The "Rehbinder effect"[134] refers to the change induced by surface-active agents, e.g. oleic acid or cetyl alcohol in paraffin. The flow of tin, copper and lead was increased in the presence of such materials[134]. The effect varied with the concentration. As this was raised, the yield strength decreased and the creep rate increased, but then a reversal occurred at concentrations thought to correspond to the formation of a monolayer. The effect was not the same magnitude for molecules of different length and varied with both the temperature of testing and the strain rate.

The interpretation of these results and of other experiments in aqueous solutions under conditions where an oxide film could not form was a physical one concerned with adsorption causing micro-crack formation.

Other workers interpret the increase in flow as a direct result of chemical attack removing the oxide, although it has been reasonably argued that the reduction in strength caused by surface-active substances is much greater than the strengthening produced by an oxide film. There may be more than one effect at work in these observations. If the reaction between a metal and its environment produces an adherent film, then the creep strength and work hardening rate are increased.

More recent work suggests that the polar molecule sare initilaly physically adsorbed and that they react chemically to form metal soaps when plastic deformation occurs[135]. The removal of metal, presumably from dislocation pile-ups, produces a lowering in work hardening. Removal of dislocations will also increase the creep rate. Electrolytic action has the same effect and increases in anodic current density produce sudden elongation, interpreted in terms of a dislocation 'pop-out' mechanism[135].

On polycrystalline materials similar effects are noted, although much of the work has been concerned with studies of fatigue and creep rupture tests. Surface-active agents appear to decrease the elongation to rupture and lower the fatigue life.

The locking of surface dislocation sources or the blocking of internally produced dislocations by a surface film are both possible mechanisms to explain most of the 'strengthening' effects already noted.

Dezincification, Impingement Attack and Cavitation

Dezincification occurs in brasses. The surface of single-phase α-brass containing 70% copper and 30% zinc gradually becomes denuded of zinc. Since this attack can occur in depth, the mechanism probably includes dissolution of the alloy followed by re-deposition of the copper. Once some of the surface atoms of zinc are removed, the remaining copper atoms will provide an efficient cathode surface and stimulate attack on the brass.

The effect produces a surface containing a 'sponge' of copper which can be pierced with a matchstick since it has practically no mechanical strength whatsoever. Brass tubes can fail rapidly in flowing environments, particularly sea water. The trouble has been largely overcome in single-phase brasses by the addition of small amounts of arsenic (0·02–0·06%). The effect is exacerbated by chloride ions, decreasing pH and high temperature. In two-phase brasses, the β-phase, which is richer in zinc, is still liable to suffer this form of attack even if arsenic is added.

Impingement attack occurs in marine condensers on the inside of tubes containing flowing environments. It occurs wherever large bubbles in the turbulent water burst on the tube surface and it produces large horseshoe-shaped pits. At the points of impingement of the bubbles there is considerable depolarization of both the anodic and cathodic reactions since the corrosion products are removed and there is an abundant supply of oxygen. The hammering of the oxide film by a constant stream of bubbles also enhances attack.

Apart from designing condensers which will produce only small bubbles which are safe, the problem has been largely overcome by using specific alloys. Since these are expensive, their safe use is associated with certain maximum water speeds. The best alloy is 70 Cu–30 Ni and this stands up to the highest water velocities required. It is employed in warships and luxury liners where a minimum weight of tubing is required. Bronzes containing 10–12·5% tin are also believed to give good resistance. Single-phase brasses containing 2% aluminium are also in widespread use. Cupro-nickel alloys also contain iron and manganese which improve pitting resistance (1% of each) and erosion resistance (2% of each), though the latter alloy may suffer pitting. Cheaper cupro-nickel alloys have been developed containing 5% and 10% nickel.

In designing marine condensers for coastal power stations, samples of local water collected at different states of the tide are tested for their aggressiveness towards a selection of these alloys. The jet impingement apparatus provides a simple device for examining the effect of variation in water speed and bubble size upon impingement resistance to the local water. After exhaustive tests a choice is made.

Cavitation attack, which occurs on ships' propellers and steam turbine blades, produces deep fissures and pits. It occurs on those parts of the blade where the hydrodynamic flow produces a localized low pressure, which in turn leads to low pressure bubbles being formed in the water. This 'milkiness', which can be seen in waters containing oxygen, does not occur in de-aerated waters. The collapse of these bubbles on or near the metal surface produces a strong compression wave. The phenomenon is the result of a combination of both a mechanical and a chemical effect. It has been seen on non-metallic materials which would have no chemical reaction with water, e.g. bakelite. Both design factors and the use of chemically resistant alloys and cathodic protection are used to combat it.

Condenser tubes are also susceptible to pitting attack caused by the settling of solid particles or marine organisms on the tube surface. The occlusion of oxygen and the dissolution of copper in the form of complexes creates a differential aeration cell. In 70/30 α-brasses the addition of 1% Sn, known as Admiralty Brass, 70–29–1, produces an alloy that resists this type of attack.

Liquid Metal and Fused Salt Corrosion

Liquid metal corrosion depends upon the speed of the metal and temperature variation throughout the system, in addition to other common factors. It may remove one element, e.g. molten bismuth and lithium both remove nickel from stainless steels, or it may penetrate down grain boundaries, e.g. mercury cracking of α-brass. Thermal metal transfer may occur from hot regions to colder regions of lower solubility for the solute metal. Metal transfer can occur also where there is no temperature gradient if an activity gradient is established. The exact attack that occurs is dependent upon many factors and much remains to be done on the subject. The mutual solubilities of the two metals, or of the two with a third, the presence of impurities, the formation of intermetallic compounds, diffusion rates of the types of species present, are all important factors.

Fused salt attack occurs if the metal is soluble in the melt, or if it is oxidized to metal ions. Noble metals tend to be soluble in molten alkali chlorides, but otherwise metals are generally soluble only in their own salts. Little information is available here. As in liquid metal corrosion, thermal metal transfer may occur with removal from the high temperature region and deposition at the cold temperature region. In addition to a temperature solubility effect, it appears that there is a faradaic mass transfer effect since an electrolytic current is required[136].

Corrosion by fused salts is lowered by reducing the oxidizing power of the melt, using alkalis which increase the activity of the protective O^{2-} species, or reactive melts that will remove oxidizing species, e.g. Cl^-, using magnesium or silicon (which form volatile chlorides).

Grain Boundary Corrosion

A grain boundary in a pure metal is more prone to be attacked than the grain itself. Use is made of this when metals are etched. In a grain boundary atoms are loosely packed, unlike atoms in the regular lattice position of the bulk material. Since boundaries have a surface energy it is not unexpected that they should dissolve at a higher rate than bulk material.

In metals in everyday use other phenomena occur in association with grain boundaries which produce corrosion reactions in addition to those expected from the intrinsic nature of boundaries.

Impurities play an important role in grain boundary attack since they are frequently found to be segregated there to some degree. Attack occurs if the grain boundary material is anodic to the grains or if it provides a surface with low overvoltage for the cathodic reaction, e.g. hydrogen evolution, when the grain edge adjacent to this grain boundary material will be attacked, as in the case of iron in aluminium, which has already been cited. Precipitates forming on grain boundaries are also important, particularly if they create local cells, as in the case of weld decay.

Aluminium alloys and austenitic stainless steels are both prone to grain boundary attack and they must be given careful heat treatment in order to minimize it. Many other materials exhibit varying degrees of grain boundary attack, which together with an applied tensile stress can result in very rapid failures. Intergranular cracking generally arises from the presence of grain boundaries of active material, but it may be determined by breakdown of the covering oxide film. Thus the stress corrosion cracking of α-brass in ammonia is intragranular at low pH where the superficial film is dissolved and intergranular at neutral pH where the film is stable.

Fretting Corrosion

Fretting corrosion is caused by two surfaces rubbing together. Corrosion product is continually removed. It can occur in anything that is vibrating. The continual wastage of metal results in severe wear and may even lead to complete failure. Fretting is often associated with fatigue failures.

On ferrous materials the brown oxide powder that forms is often referred to as 'cocoa'. The amount produced decreases with increasing humidity and increasing temperature, although this latter point has not been examined much. In inert atmospheres iron particles are produced and at low pressures actual metal seizure (welding) may occur. Soft metals generally suffer more than hard metals.

The mechanism of fretting corrosion has not been fully established. Initially, when the metals make contact, the proud parts of the surfaces weld together and the oscillatory movement produces adhesive wear, since tearing occurs. Both surfaces oxidize and any loose debris causes abrasive wear between the rubbing surfaces which are continually re-oxidized so that the surface is gradually worn away. Moisture tends to reduce fretting, probably by acting as a lubricant, while an increase in temperature can often be beneficial, for reasons that have not been investigated.

A phosphated surface containing oil, common lubricants and soft metal deposits all help reduce fretting corrosion where surfaces are intended to rub. Otherwise, mechanical tightening can be applied to prevent rubbing together.

4.8. CORROSION TESTING

The range of experimental techniques that are used in corrosion studies is considerable. The data help towards the determination of corrosion mechanisms which in turn should at least distinguish between possible and impossible protective measures and at best the most suitable method of reducing corrosion to an acceptable level. In practice it is frequently extremely difficult to interpret data. Often there is not time to gather sufficient of it, and, in general, laboratory results do not always correlate with practical experience. All these factors and others lead to the practice of corrosion testing, which is a methodical and not necessarily empirical way of determining how a metal behaves in specific atmospheres that it is likely to be exposed to.

The simplest and most obvious method of corrosion testing consists of putting the metal, either in a semi-finished or fabricated form, in the environment and examining it from time to time. Most large metal producers and users have extensive programmes of this type which consist largely of investigating the effects of atmospheric and marine conditions of exposure. It is the most reliable type of test since it reproduces exactly the service conditions. There are several drawbacks. Firstly, it can take a very long time and for that reason it is impractical in many circumstances. Secondly, it may be uneconomic to do, and this objection covers a very wide range of reasons.

In order to obtain data much more quickly than by simple exposure, accelerated testing procedures are widely used. Many laboratories, for example, use a salt spray room where specimens can be exposed to specific conditions of salinity, humidity and temperature. Behaviour under accelerated conditions does not always correspond to that under exposure conditions and care must be exercised here. Laboratory environments are not always reliable substitutes for natural ones. It has already been noted (cf. section 2.2) that 3% NaCl is slightly more aggressive than sea water of the same salinity since the latter contains inhibitive magnesium ions. Should the sea water environment contain cysteine (a compound derived from sea-weed which depolarizes the hydrogen evolution reaction), then clearly 3% NaCl solution will not give the same results. The differences here depend upon the specific conditions; they may be quite important.

Corrosion testing programmes require considerable care in preparation and a sound understanding of the variables involved; e.g. in atmospheric testing, specimens that are horizontal will be wetter for longer than specimens that are inclined. Testing bridges the gap between laboratory experiments and actual usage. Every corrosion advance requires a prolonged testing appraisal, be it a new alloy for oxidation resistance or the technique of anodic protection, and it is the thoroughness and care with which this is done that often determines progress in corrosion prevention.

There are specific tests, specific pieces of apparatus specially designed for some of them, and special techniques, all devised to give reliable data on corrosion resistance. A few points can be made to indicate the extent and difficulties of the whole subject.

Specimen preparation must be standardized. Oxygen content, environmental temperature and velocity must be controlled. Statistical methods have been very successfully used, but a good knowledge of the subject is required. In pitting studies, for example, if the probability of incidence is low, small specimens may fail to reveal any. If the metal is to be used in large sheets, then perforation may result from the occurrence of a single pit, where tests showed the metal to be immune. In stress corrosion tests, U-bend specimens often give different results from uniaxial tensile test specimens since, in the latter, specimens are exposed to increasing stress. Differences in time to failure may give quite false information about stress corrosion susceptibility if the thickness of the initial oxide film is not the same on all specimens, since the breakdown of the oxide film may take very much longer than the time of crack propagation. Slight differences in pH of the testing medium may also cause spurious results, since the oxide film may be removed at very different rates over a narrow pH range.

4.9. LAST WORDS

Such a short account of corrosion, which is a wide ranging subject, must necessarily contain many simplifications while also imposing severe limitation on the choice of topics. In this final section it is intended to underline some of the complexities of several topics already mentioned and to include short accounts of specific matters that the author finds of special interest.

Reduction Processes

It is usual to consider that the common cathodic reaction of oxygen reduction occurs as

$$O_2 + 2H_2O + 4 \text{ electrons} \rightarrow 4\,OH^-.$$
$$\text{(solution)}$$

However, an alternative reaction is possible, viz.

$$2\,O_2 + 4\,H^+ + 4 \text{ electrons} \rightarrow 4\,OH^-.$$
$$\text{(solution)}$$

Written in that form it is easy to see that the variation of potential with pH will be the same for both reactions and that therefore the Pourbaix diagram does not distinguish between them. Concentration effects are the same for both reactions. The first reaction is the most likely in neutral solutions, where the concentration of hydrogen ions will be very low (10^{-7}), while the second will become increasingly important as the pH of the solution falls and hydrogen ions become available. Since these are adsorbed on to the cathode surface there is also the possibility of

$$H^+_{(adsorbed)} + H^+_{(adsorbed)} + 2 \text{ electrons} \rightarrow H_2 \text{ (evolved).}$$

Since oxygen solubility generally falls off as pH is lowered, this reaction gradually becomes the predominant one. Thus the second reaction concerning oxygen reduction is only likely to occur at mildly acidic values of pH. The assumption throughout is that these reactions occur on an ideal homogeneous metal surface. Commonly heterogeneities will severely upset this picture, by producing, for example, local action cells and if phases occur which have very low hydrogen overpotential, then hydrogen evolution may readily occur in neutral solutions.

Oxygen reduction occurs in stages. Hydrogen peroxide is often an intermediate product:

$$O_2 + 2\,H_2O + 2 \text{ electrons} \rightarrow H_2O_2 + 2\,OH^-.$$

It is easily reduced and therefore not always readily detectable. It is often used as an addition in etching solutions, since it provides oxygen and therefore helps to speed up reactions, by depolarizing the cathodic reaction.

A further possibility is

$$4\,H^+ + O_2 + 4\,\text{electrons} \rightarrow 2\,H_2O$$

which can also occur in mildly acidic solutions.

Apart from consideration of overall pH, the occurrence of these different reactions depends upon the value of the local pH developed at corrosion sites, particularly under stagnant conditions. Furthermore, the ease of adsorption of the charged species, which is partly determined by the nature of the surface and partly by the alloy characteristics of the electrode, influences which reaction occurs.

The reduction of nitric acid has been referred to in several sections. Several different reduction products are found and a total analysis of the several reactions would be extremely complicated. It was explained in section 2.8 that concentrated nitric acid renders iron passive, while in dilute acid iron corrodes very readily. This is only a simple picture. The attack upon iron by nitric acid is catalysed by nitrogen peroxide, which is a reaction product, and therefore the mechanism is autocatalytic. For noble metals, e.g. copper, unless the nitrogen peroxide can accumulate at the metal surface, attack will not occur. Thus copper and silver specimens can be rotated in dilute nitric acid without undergoing attack.

Local Cathodes

In section 2.4 hydrogen ion reduction has been described as one of the principal cathodic reactions that occurs when metals corrode in aqueous solutions. While hydrogen evolution can be anticipated when a metal is polarized cathodically, it is sometimes not always immediately understood that hydrogen evolution can occur under conditions of anodic polarization, yet such observations are widespread in localized corrosion processes, such as have been discussed in this chapter, pitting, stress corrosion cracking and corrosion fatigue. Such local cathodes can arise from potential differences existing between the outer surface of a specimen and the inner occluded cell. It is also commonly associated with passive metals. In both cases, the observed changes in pH—often but not always acidification—will also be important. Two different cases can be described very simply, although there can be many complexities.

If pitting is induced by anodic polarization, hydrogen evolution will occur if the pitting or breakdown potential, E_b, is active with respect to the hydrogen evolution potential at that pH. Pitting on aluminium, for example, occurs at a potential relatively close to -550 mV E_H, while acidity at pH 3·5 within a pit would promote hydrogen evolution at a potential of $-3·5 \times 60$, viz. -210 mV E_H. Even allowing for some additional overpotential, it is reasonable that hydrogen evolution should occur within pits and cracks in aluminium alloys. That hydrogen evolution does not occur on the filmed surface to any significant degree is due to the poor conductivity of the film. As the potential of the specimen is raised above E_b, the pitting current will increase and the amount of hydrogen evolved will increase at the same time. At more noble potentials a greater pit density may be associated with even larger amounts of hydrogen evolution, even though the potential difference causing hydrogen evolution is diminishing as the potential of the specimen becomes more noble and approaches E_H for the solution with the pH developed at the pit surface. When hydrogen evolution stops will depend upon the iR drop of the solution and any film down the pit. If this is large, then raising the potential of the outer surface of a specimen to values noble to the hydrogen evolution potential may not be sufficient to ensure that the potential at the tip of the pit or crack is sufficiently noble to exclude hydrogen evolution. In extreme cases, e.g. titanium, hydrogen evolution occurs within pits on specimens polarized to $+2V E_H$.

Anodic Reactions

In section 2.8 some attempt was made to describe the state of passivity and to consider it as the alternative to active dissolution. While these two reactions are of basic importance, they are not the only two possibilities. Under some circumstances, for example, levelling and brightening of the surface may occur or oxidation of the metal to a higher valency species of some solubility. At a metal anode a number of reactions is possible and an understanding of corrosion requires an appreciation of the factors that determine which one occurs. The different general types of anodic polarization curves that can occur[137] are shown in Fig. 88. The metal may exhibit active dissolution or

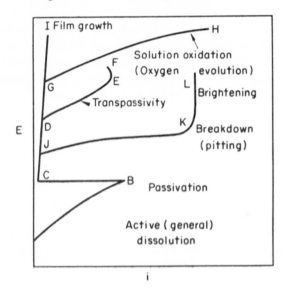

Fig. 88. Schematic diagram of various anodic polarization curves that may arise in a number of different environments and with a range of alloys. Notice that current rather than the logarithmic of the current density is drawn as the abscissa. A small increase in $i_{passive}$ as the potential becomes more noble is shown which may not always be found[137].

passivity. In certain media passivity will break down to give pitting at E_b as already discussed. At noble potentials the bottom of the pits may be brightened. If a soluble species is formed, e.g. CrO_4^- on stainless steel surfaces, a region of transpassivity is observed which may be succeeded at a higher potential by a region of secondary passivity. Provided that the film is sufficiently conductive, some species in the solution may be oxidized, e.g. oxygen evolution. With films possessing a low electronic conductivity, application of an anodic polarization causes film thickening, as described briefly in section 3.6 on anodizing. With so many possible anodic reactions it is not surprising that confusion can arise, particularly since more than one process may be occurring at a particular potential.

The process of passivity has been considered in detail by Hoar[137]. Active dissolution of metal occurs from kink sites which occupy only a very low proportion of the total metal surface. Passivity is associated with the adsorption of water molecules on to the metal, a process that can occur all over the surface. A subsequent loss of protons from the adsorbed water molecules is accompanied by a simultaneous uptake of any surface metal ions by O^{2-} or OH^- into positions between them. This is regarded as the primary act of passivation and appears to reconcile different concepts based upon an adsorbed layer and a three-dimensional oxide film. On this basis, passivity replaces active dissolution when it becomes kinetically the easier process. The brightening process arises from having a layer in contact with the metal surface, which may be a viscous liquid or a salt film, consisting of the metal ions and some anions, neither containing many water molecules. The current through the layer depends upon the thickness, which is smallest over protuberances which therefore dissolve more rapidly so that levelling of the surface gradually occurs.

The presence in the electrolyte of negative species will affect the passivation process to an extent dependent upon the nature of the species and its concentration. Readily adsorbable species like Cl^- in competing with the adsorption of the water molecules $\overset{-}{O}\!\!\diagup\!\!\overset{\displaystyle H^+}{\diagdown H}$ may prevent complete passivity from being established. Once it is established the current density is dependent upon the movement of cations into the film, across the film and into the electrolyte.

The two factors that most closely determine anodic behaviour are the potential and the ratio of anion/water molecules adjacent to the metal surface, a ratio that may differ considerably from that in the bulk

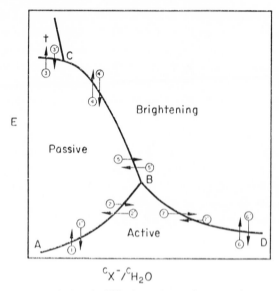

Fig. 89. Schematic drawing[137] of anodic reactions at various potentials and in solutions of various anion/water concentration ratios. 1, anodic passivation; 1′, activation at the Flade potential; 2, activation by adding an aggressive anion; 2′, inhibition achieved by removing or suppressing an aggressive anion; 3, transpassivation; 3′, repassivation from transpassivation achieved by lowering the anode potential; 4, breakdown by pitting, which is followed by brightening if the anode potential is raised; 4′, repassivation by lowering the anode potential; 5, breakdown by pitting, leading to brightening, as a result of adding an aggressive anion; 5′, repassivation by removing or suppressing an aggressive anion; 6, anodic brightening, by raising the anode potential in concentrated solution; 6′, anodic etching of brightening metal achieved by lowering potential in concentrated solution; 7, anodic brightening, from etching, by increasing the anion concentration; 7′, anodic etching, from brightening, by decreasing the anion concentration.

of the solution. The effect[137] of these two variables upon a wide range of possible reactions is shown in Fig. 89. The diagram is divided into four regions, the relative extents of which will depend upon the metal and the anion involved. These are passive, active, brightening and transpassive. Not all four will always be found. Transpassivity is, for example, limited to a relatively few metals. The values of the potentials at which transitions occur will depend upon the relative nobility of the particular metal. The numbers refer to pairs of reactions induced by altering the potential or the anion/water molecule ratio. No attempt is made to consider the complexities arising from altering both at the same time.

The diagram is not exclusive and other phenomena may occur. Each region may be capable of subdivision, e.g. there may be more than one type of passive film. Different anions will have different effects. Furthermore, there is no indication of the relative kinetics of these reactions. Figure 89 is therefore an essentially simple representation which may require alteration when it is further investigated. Nevertheless, it does serve well to relate the various anodic processes that a metal may undergo.

Bibliography

This bibliography is divided into four sections:

1. Introductory books on corrosion.
2. General books on corrosion, which may be suitable for reference as well as for acquiring a deeper understanding than is provided by books in category 1.
3. Books on a special corrosion topic, which in many cases represent the most advanced state of knowledge in book form.
4. Books on pure science which will amplify subjects that have been lightly touched on in this book. Since all these books deal with a wide range of topics, the specific subject is indicated in parentheses.

1. INTRODUCTORY BOOKS

An Introduction to Metallic Corrosion, U. R. Evans, Arnold (London), 1st edition 1948, 2nd edition 1972.

Electrodeposition and Corrosion Processes, J. M. West, Van Nostrand (London), 1st edition 1965, 2nd edition 1972.

Principles of Metallic Corrosion, J. P. Chilton, Royal Institute of Chemistry (London), 1st edition 1961, revised edition 1962, 2nd edition 1969.

2. GENERAL BOOKS

Corrosion Handbook, Ed. H. H. Uhlig, Wiley (New York), 1948.

Corrosion, Ed. L. L. Shreir (2 volumes), Newnes (London), 1963.

Metallic Corrosion, Passivity and Protection, U. R. Evans, Arnold (London), 2nd edition 1947.

The Corrosion and Oxidation of Metals, U. R. Evans, Arnold (London), 1960.

The Corrosion and Oxidation of Metals: first supplementary volume, U. R. Evans, Arnold (London), 1968.

Corrosion and Corrosion Control, H. H. Uhlig, Wiley (New York), 1st edition 1963, 2nd edition 1971.

Lectures on Electrochemical Corrosion, M. Pourbaix, Plenum Press (New York), 1972.

Corrosion: Causes and Prevention, F. N. Speller, McGraw-Hill (New York), 1951.
Corrosion Resistance of Metals and Alloys, F. L. LaQue and H. R. Copson, Reinhold (New York), 1962.
Corrosion Engineering, M. G. Fontana and N. D. Greene, McGraw-Hill (New York), 1967.
Corrosion and Its Prevention in Waters, G. Butler and H. C. K. Ison, Leonard Hill (London), 1966.
The Theory of Corrosion and Protection of Metals, N. D. Tomashov, Macmillan (London), 1966.
Passivity and Protection of Metals against Corrosion, N. D. Tomashov, Plenum Press (New York), 1967.

3. BOOKS ON SPECIFIC TOPICS

Oxidation of Metals and Alloys, O. Kubaschewski and B. E. Hopkins, Butterworths (London), 1962.
High Temperature Oxidation of Metals, P. Kofstad, Wiley (New York), 1966.
Oxidation of Metals and Alloys, K. Hauffe, Plenum Press (New York), 1965.
The Rusting of Iron: Causes and Control, U. R. Evans, Arnold (London), 1972.
Cathodic Protection, J. H. Morgan, Leonard Hill (London), 1959.
Corrosion Inhibitors, J. I. Bregman, Macmillan (New York), 1963.
Metallic Corrosion Inhibitors, I. N. Putilova, S. A. Balezin, V. P. Barannik. Trans. G. Ryback, Pergamon Press (Oxford), 1960.
Protective Coatings for Metals, R. M. Burns and W. W. Bradley, Reinhold (New York), 2nd edition 1955.
Anodic Oxide Films, L. Young, Academic Press (London), 1961.
The Surface Treatment and Finishing of Aluminium and Its Alloys, S. Wernick and R. Pinner, Draper (London), 1959.
The Principles of Metal Surface Treatment and Protection, D. R. Gabe, Pergamon (London), 1972.
Potentiostat and its Applications, J. A. von Fraunhofer and C. H. Banks, Butterworths (London), 1972.
The Corrosion of Light Metals, H. P. Godard, W. B. Jepson, M. R. Bothwell and R. L. Kane, Wiley (New York), 1967 (covers aluminium, magnesium, beryllium and titanium).
The Corrosion of Copper, Tin and Their Alloys, H. Leidheiser, Wiley (New York), 1971.
Marine Corrosion, T. H. Rogers, Newnes (London), 1968.
Marine Corrosion Handbook, T. H. Rogers, McGraw-Hill (New York), 1960.
Fundamental Aspects of Stress Corrosion Cracking, Ed. R. W. Staehle, A. J. Forty and D. van Rooyen, NACE (Houston), 1969.
The Theory of Stress Corrosion Cracking in Alloys, Ed. J. C. Scully, N.A.T.O. (Brussels), 1971.
Stress Corrosion Cracking in High Strength Steels and in Titanium and Aluminium Alloys, Ed. B. F. Brown, Naval Research Laboratory (Washington, D.C.), 1972.

Corrosion Fatigue, Ed. O. Devereux, A. J. McEvily Jr., and R. W. Staehle, N.A.C.E. (Houston), 1972.

Fretting Corrosion, R. B. Waterhouse, Pergamon (London), 1972.

Progress in Materials Science, Vol. 9, No. 3, 1961. *Effects of Environment on Mechanical Properties of Metals*, I. Kramer and L. J. Demer, Pergamon Press (Oxford).

Effect of a Surface-active Medium on the Deformation of Metals, V. I. Likhtman, P. A. Rehbinder, G. V. Karpenko, H.M.S.O. (London), 1958 (Translated from Russian).

Corrosion—Mechanical Strength of Metals, L. A. Glikman. Translated by J. S. Shapiro, Butterworths (London), 1961.

4. BOOKS ON PURE SCIENCE

Introduction to Solid State Physics, C. Kittel, Wiley (New York), 1956. (Semiconductors.)

Physical Chemistry, W. J. Moore, Longmans, Green (London), 1957. (Chemical thermodynamics.)

Metallurgical Thermochemistry, O. Kubaschewski and E. Ll. Evans, Pergamon Press (Oxford), 1958. (Free energy.)

Electrochemistry, Principles and Applications, E. C. Potter, Cleaver-Hulme (London), 1959.

References

1. R. C. EVANS, *The Introduction to Crystal Chemistry*, Cambridge University Press, Cambridge, p. 217 (1952).
2. C. J. SMITHELLS, *Metals Reference Book*, Butterworths, London, p. 590 (1955).
3. D. W. HOPKINS, *Physical Chemistry and Metal Extraction*, J. Garnet Miller, London, p. 112 (1954).
4. J. BERNARD, F. GRØNLUND, J. ONDAR and M. DURET, *Z. Electrochem.*, **63**, 799 (1959).
5. M. D. SANDERSON and J. C. SCULLY, *Corrosion*, **25**, 291 (1969).
6. P. T. LANDBERG, *J. Chem. Phys.*, **23**, 1079 (1955).
7. H. H. UHLIG, *Corros. Sci.*, **7**, 235 (1967).
8. A. MacRAE, *Surface Sci.*, **1**, 319 (1964).
9. O. KUBASCHEWSKI and B. E. HOPKINS, *Oxidation of Metals and Alloys*, Butterworths, London, p. 78 (1962).
10. U. R. EVANS, *Trans. Electrochem Soc.*, **91**, 547 (1947).
11. M. H. DAVIES, M. T. SIMNAD, and C. E. BIRCHENALL, *J. Metals*, **3**, 889 (1951); **5**, 1250 (1953).
12. G. VALENSI, *Rev. Metall.*, **45**, 10 (1948).
13. G. W. CASTELLAN and W. J. MOORE, *J. Chem. Phys.*, 1741 (1949).
14. C. WAGNER, *Z. Phys. Chem.* (*B*), **21**, 25 (1933).
15. T. P. HOAR and L. E. PRICE, *Trans. Faraday Soc.*, **34**, 867 (1938).
16. C. WAGNER, *Atom Movements*, American Society for Metals, Cleveland, p. 153 (1951).
17. Ref. 9, p. 128.
18. U. R. EVANS, *Internal Stresses in Metals and Alloys*, Institute of Metals, London, p. 291 (1947).
19. N. B. PILLING and R. E. BEDWORTH, *J. Inst. Met.*, **29**, 529 (1923).
20. H. DÜNWALD and C. WAGNER, *Z. Phys. Chem.* (*B*), **22**, 212 (1938).
21. J. GUNDERMANN, K. HAUFFE and C. WAGNER, *Z. Phys. Chem.* (*B*), **37**, 148 (1937).
22. C. WAGNER and K. GRÜNEWALD, *Z. Phys. Chem.*, **40**, 455 (1938).
23. C. A. HOGARTH, *Z. Phys. Chem.*, **198**, 30 (1951).
24. D. P. WHITTLE, D. J. EVANS, D. B. SCULLY and G. C. WOOD, *Acta Met.*, **15**, 1421 (1967).
25. M. D. SANDERSON and J. C. SCULLY, *Oxidation of Metals*, **3**, 59 (1971).

26. J. P. DENNISON and A. PREECE, *J. Inst. Met.*, **81**, 229 (1953).
27. L. DE BROUCHÈRE and L. HULRECHT, *Bull. Soc. Chim. Belg.*, **60**, 311 (1951).
28. L. DE BROUCHÈRE and L. HULRECHT, *Bull. Soc. Chim. Belg.*, **61**, 101 and 205 (1952).
29. A. M. PORTEVIN, E. PRETEL and H. JOLIVET, *J. Iron Steel Inst.*, **103**, 219 (1934).
30. C. WAGNER, *J. Electrochem. Soc.*, **99**, 369 (1952).
31. F. N. RHINES, *Trans. Amer. Inst. Min. (Metall.) Eng.*, **137**, 246 (1940).
32. L. E. PRICE and G. J. THOMAS, *J. Inst. Met.*, **63**, 21 and 29 (1938).
33. M. D. SANDERSON and J. C. SCULLY, *Corros. Sci.*, **10**, 165 (1970).
34. M. HANSEN, *The Constitution of Binary Alloys*, McGraw-Hill, New York, p. 676.
35. J. PAIDASSI, *Acta Met.*, **6**, 184 (1958).
36. K. HAUFFE and H. PFEIFFER, *Z. Metallk.*, **44**, 27 (1953).
37. G. C. WOOD and M. G. HOBBY, *J. Iron Steel Inst.*, **203**, 54 (1965).
38. N. CABRERA and N. F. MOTT, *Rep. Prog. Phys.*, **12**, 163 (1943-9).
39. C. N. COCHRAN and W. C. SLEPPY, *J. Electrochem. Soc.*, **108**, 322 (1961).
40. Ref. 9, p. 106.
41. U. R. EVANS, *The Corrosion and Oxidation of Metals*, Arnold, London, p. 789 (1960).
42. U. R. EVANS and L. C. BANNISTER, *Proc. Roy. Soc. (A)*, **125**, 370 (1929).
43. L. B. PFEIL, *J. Iron Steel Inst.*, **119**, 501 (1929) and **123**, 237 (1931).
44. U. R. EVANS, *Metal Ind. (London)*, **29**, 481 (1926).
45. U. R. EVANS and T. P. HOAR, *Proc. Roy. Soc. (A)*, **137**, 343 (1932).
46. T. P. HOAR, The Anodic Behaviour of Metals, *Modern Aspects of Electrochemistry*, No. 2, Ed. J. O'M. Bockris, Butterworths, London, p. 262 (1959).
47. M. POURBAIX, *Atlas d'Equilibres Electrochemiques à 25 °C.*, Gauthier-Villars, Paris, p. 312 (1963).
48. E. C. POTTER, *Electrochemistry*, Cleaver-Hulme, London, p. 80 (1961).
49. M. POURBAIX, *Thermodynamics of Aqueous Solutions* (trans. J. N. Agar), Arnold, London (1949).
50. M. POURBAIX, *Corrosion*, **26**, 431 (1970).
51. M. STERN, *J. Electrochem. Soc.*, **102**, 609, 663 (1955).
52. U. R. EVANS, ref. 34, p. 878.
53. R. VONDRACEK and IZAK KNZKO, *J. Trans. Chim. Pays-Bas*, **44**, 376 (1925).
54. C. W. BORGMANN and U. R. EVANS, *Trans. Electrochem. Soc.*, **65**, 249 (1934).
55. J. O'M. BOCKRIS and E. C. POTTER, *J. Electrochem. Soc.*, **99**, 169 (1952).
56. J. BUMBULIS and W. F. GRAYDON, *J. Electrochem. Soc.*, **109**, 1130 (1962).
57. W. A. MUELLER, *Corrosion*, **18**, 734 (1962).
58. M. STERN, *Corrosion*, **14**, 329t (1958).
59. R. F. STEIGERWALD and N. D. GREENE, *J. Electrochem. Soc.*, **109**, 1026 (1962).
60. A. JOSHI and D. F. STEIN, *Corrosion*, **28**, 321 (1972).
61. C. EDELEANU and T. LAW, *Phil. Mag.*, **7**, 573 (1962).
62. P. LACOMBE and L. BEAUJARD, *J. Inst. Metals*, **74**, 1 (1948).
63. C. EDELEANU, *J. Inst. Met.*, **89**, 90 (1061).

64. F. P. Bowden and N. R. Throssell, *Proc. Roy. Soc. (A)*, **209**, 297 (1951).
65. G. Schikorr, *Werkstoffe Korros.*, **14**, 69 (1963); **12**, 457 (1964); **18**, 514 (1967).
66. U. R. Evans and C. A. Taylor, *Corros. Sci.*, **12**, 227 (1972).
67. J. C. Hudson, *The Corrosion of Iron and Steel*, Chapman & Hall, London, p. 66 (1940).
68. N. D. Greene, *Corrosion*, **15**, 369t (1959).
69. C. Edeleanu, *Chem. Ind.*, 301 (1961).
70. M. J. Pryor, *J. Electrochem. Soc.*, **106**, 557 (1959).
71. R. P. Frankenthal, *J. Electrochem. Soc.*, **114**, 542 (1967).
72. H. H. Uhlig, *Corrosion and Corrosion Control*, McGraw-Hill, New York, p. 60 (1963).
73. P. F. King and H. H. Uhlig, *J. Phys. Chem.*, **63**, 2026 (1959).
74. *Corrosion Resistance of Titanium*, IMI (Kynoch) Ltd., Birmingham, p. 60 (1970).
75. N. D. Greene, C. R. Bishop and M. Stern, *J. Electrochem. Soc.*, **108**, 836 (1961).
76. N. D. Tomashov, *Corrosion*, **14**, 229t (1958).
77. E. L. Chappell, *Ind. Eng. Chem.*, **22**, 1204 (1930).
78. U. R. Evans, ref. 34, p. 927.
79. P. Kofstad, *High Temperature Oxidation of Metals*, Wiley, London, p. 300 (1966).
80. J. D. Sudbury, O. L. Riggs and D. A. Shock, *Corrosion* **16**, 47t (1960).
81. U. R. Evans and D. E. Davies, *J. Chem. Soc.*, p. 2607 (1951).
82. R. B. Mears and U. R. Evans, *Proc. Roy. Soc. (A)*, **146**, 153 (1934).
83. R. B. Mears and U. R. Evans, *Trans. Faraday Soc.*, **31**, 527 (1935).
84. M. Cohen, *Corrosion*, Ed. L. L. Shreir, Newnes, London pp. 18, 27 (1963).
85. *Report of the National Chemical Laboratory, 1958*, H.M.S.O. London, p. 28 (1959).
86. M. J. Pryor and M. Cohen, *J. Electrochem. Soc.*, **100**, 203 (1953).
87. D. M. Brasher, *Nature, London*, **185**, 838 (1960).
88. M. Stern, *J. Electrochem. Soc.*, **105**, 638 (1958).
89. G. Butler and H. C. K. Ison, *Corrosion and its Prevention in Waters*, Leonard Hill, London, p. 27 (1966).
90. *Corrosion Technology*, May, p. 42 (1964).
91. A. C. Makrides and M. Stern, *J. Electrochem. Soc.*, **107**, 877 (1960).
92. T. P. Hoar and R. D. Holliday, *J. Appl. Chem.*, **3**, 502 (1953).
93. S. Glasstone, *An Introduction to Electrochemistry*, Van Nostrand, New York, p. 428 (1924).
94. *Ukrain. Khim. Zhurn.*, **58(9)**, 1066 (1963).
95. J. E. O. Mayne and D. van Rooyen, *J. Appl. Chem.*, **4**, 384 (1954).
96. B. W. Cherry and J. E. O. Mayne, *First International Congress on Metallic Corrosion*, Butterworths, London, p. 539 (1962).
97. T. K. Ross, *Trans. Inst. Chem. Eng.* **40**, 83 (1962).
98. *Occluded Cell Corrosion*, NACE, Houston, Texas. To be published.
99. D. Vermilyea and C. Tedman Jr., *J. Electrochem. Soc.*, **121**, 321 (1972).

230   References

100. Z. SZKLARSKA-SMIALOWSKA, *Corrosion*, **27**, 223 (1971). –
101. Z. SZKLARSKA-SMIALOWSKA and M. JANIK-ZICHOR, *Brit. Corros. J.*, **4**, 138 (1969).
102. E. N. PUGH, J. V. CRAIG and W. G. MONTAGUE, *Trans. ASM*, **61**, 468 (1968).
103. T. FLIS and J. C. SCULLY, *Corrosion*, **24**, 326 (1968).
104. T. P. HOAR and J. G. HINES, *J. Iron Steel Inst.*, **182**, 124 (1956).
105. R. E. REED and H. W. PAXTON, ref. 96, p. 539.
106. B. F. BROWN, *Stress Corrosion Cracking in High Strength Steels, and in Titanium and Aluminium Alloys*, NRL, Washington D.C. (1972).
107. M. O. SPEIDEL, *The Theory of Stress Corrosion Cracking in Alloys*, Ed. J. C. Scully, NATO, Brussels, p. 289 (1971).
108. A. J. MCEVILY JR. and A. P. BOND, *J. Electrochem. Soc.*, **112**, 131 (1965).
109. M. J. ROBINSON and J. C. SCULLY, *Stress Corrosion Cracking and Hydrogen Embrittlement of Ferrous Alloys*, NACE, Houston. To be published.
110. H. L. LOGAN, *J. Res. NBS*, **48**, 99 (1952).
111. J. C. SCULLY, *Corros. Sci.*, **7**, 197 (1967).
112. D. T. POWELL and J. C. SCULLY, *Corros. Sci.*, **10**, 371 (1970).
113. J. A. FEENEY and M. J. BLACKBURN, ref. 107, p. 355.
114. B. F. BROWN, C. T. FUJII and E. P. DAHLBERG, *J. Electrochem. Soc.*, **116**, 218 (1969).
115. A. J. FORTY, *The Physical Metallurgy of Stress Corrosion Fracture*. Ed. T. N. Rhodin, Interscience, New York, p. 99 (1959).
116. T. P. HOAR and J. M. WEST, *Proc. Roy. Soc. (A)*, **268**, 304 (1962).
117. T. P. HOAR and J. C. SCULLY, *J. Electrochem. Soc.*, **111**, 348 (1964).
118. J. C. SCULLY and T. P. HOAR, *Second International Congress on Metallic Corrosion*, New York (1966).
119. D. TROMANS and J. NUTTING, *Fracture of Solids*, Interscience, New York, p. 637 (1963).
120. P. P. SNOWDEN, *Nuclear Eng.* Oct. (1961).
121. N. A. NEILSEN, ref. 115, p. 121.
122. J. G. HINES and E. R. W. JONES, *Corros. Sci.*, **1**, 1 (1961).
123. R. STICKLER and S. BARNARTT, *J. Electrochem. Soc.*, **109**, 343 (1962).
124. A. J. FORTY and P. HUMBLE, *Phil. Mag.*, **8**, 247 (1963).
125. R. J. GEST and A. R. TROIANO, *Hydrogen in Metals*, Paris (1972). To be published.
126. J. SPURRIER and J. C. SCULLY, *Corrosion*, **28**, 453 (1972).
127. M. O. SPEIDEL, M. J. BLACKBURN, T. R. BECK and J. A. FEENEY, *Corrosion Fatigue*, Ed. O. Devereux, A. J. McEvily Jr., and R. W. Staehle, NACE, Houston, p. 324 (1972).
128. R. M. LATANISION and A. R. C. WESTWOOD, *Advances in Corrosion Science*, Ed. R. W. Staehle and M. G. Fontana, NACE (in press).
129. A. R. C. WESTWOOD, *Strengthening Mechanisms: Metals and Ceramics*, Syracuse University Press, p. 407 (1966).
130. T. L. JOHNSTON, R. G. DAVIES and T. S. STOLOFF, *Phil. Mag.*, **12**, 305 (1965).
131. C. M. PREECE and A. R. C. WESTWOOD, *Trans. ASM*, **62**, 418 (1969).

132. F. N. RHINES, J. A. ALEXANDER and W. F. BARCLAY, *Trans. ASM*, **55**, 22 (1962).
133. S. HARPER and A. H. COTTRELL, *Proc. Phys. Soc. (B)* **63**, 331 (1950).
134. F. I. LIKHTMAN, P. A. REHBINDER and G. V. KARPENKO, *Effect of a Surface Active Medium on the Deformation of Metals*, H.M.S.O., London (1958).
135. I. KRAMER and L. J. DEMER, *Effects of Environment on Mechanical Properties of Metals*, Progress in Materials Science Series, Pergamon Press, Oxford (1961).
136. C. EDELEANU and J. G. GIBSON, *J. Inst. Met.*, **88**, 321 (1960).
137. T. P. HOAR, *Corros. Sci.*, **7**, 341 (1967).

Index